PRENTICE HALL
SCIENCE

TEACHER'S DESK REFERENCE

A Professional

Guide for

Science

Educators

Includes:

▼ Program Overview

▼ Professional Articles

▼ Classroom Resources

Prentice Hall
A Division of Simon & Schuster
Englewood Cliffs, NJ 07632

Teacher's Desk Reference
Prentice Hall Science
Second Edition

PRENTICE HALL
A Division of Simon & Schuster
Englewood Cliffs, New Jersey 07632

ISBN 0-13-402264-5

2 3 4 5 6 7 8 9 10   96 95 94

Credits begin on page 155.

TEACHER'S DESK REFERENCE

Contents

PROGRAM OVERVIEW

PROFESSIONAL ARTICLES

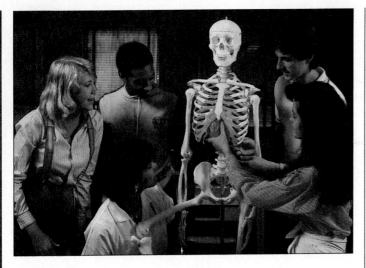

CLASSROOM RESOURCES

Contributing Writers

Dr. LaMoine L. Motz
Coordinator of Science, Health, and Outdoor Education
Oakland Public Schools
Pontiac, Michigan
President of the National Science Teachers Association for 1988–1989

Anthea Maton
former NSTA National Coordinator
Project Scope, Sequence, and Coordination
Washington, DC

Dale Rosene
Science Teacher
Marshall Middle School
Marshall, Michigan
President, National Middle Level Science Teachers Association

Dr. George Ladd
Professor of Education
Boston College
School of Education
Wooster, Massachusetts

Robert M. Jones
Professor of Curriculum and Instruction in the School of Education
University of Houston–Clear Lake
Houston, Texas

Joan Develin Coley
Chairperson of the Education Department
Western Maryland College
Westminster, Maryland

Steven J. Rakow
Associate Professor of Science Education
University of Houston–Clear Lake
Houston, Texas

Pat Hollis Smith
Former ESL Instructor
Beaumont, Texas

Dr. Karen K. Lind
Assistant Professor of Science Education
Department of Early and Middle Childhood Education
University of Louisville
Louisville, Kentucky

Amy Wagner
Coordinator of Science K–8
Brookline Public Schools
Brookline, Massachusetts

Richard Thibeault
Science Teacher
Jonas Clarke Middle School
Lexington, Massachusetts

Welcome to *Prentice Hall Science*!

The authors of *Prentice Hall Science* recognize that **education** is the predominant force in helping students understand the impact of science on their lives—and that you, the teacher, are their guide through this fascinating and often complex world. This program was therefore designed to provide you with everything you'll need to meet this challenge today and in the future.

Prentice Hall Science will help you provide your students with a basic **knowledge** of life, earth, and physical science, with special emphasis on how this knowledge relates to them and their personal range of experiences. Beyond that, the program will help you teach your students to develop an **understanding** of, and **appreciation** for, the basic concepts of science.

There are many different and exciting ways the program will accomplish these objectives. This *Teacher's Desk Reference* will acquaint you with some of the newer ones. We invite you to browse through the articles on key issues in science education . . . discover the unique integration of skills instruction and content, and the thematic strands that run throughout . . . examine the extensive teaching support and thrill to the exciting color and layout of the student pages.

We hope you will refer to this teaching aid often. Like the entire *Prentice Hall Science* program, it was developed for teachers like you by teachers like you. Please let us know how we can make future editions more helpful by calling us toll-free at 1-800-848-9500.

Thank you, and enjoy!

Jane Antoun
President

Now You Can Choose the Perfect

The new **Prentice Hall Science** *program consists of 19 hardcover books, each of which covers a particular area of science. All of the sciences are represented in the program so you can choose the perfect fit to your particular curriculum needs.*

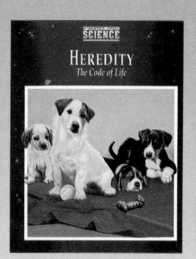

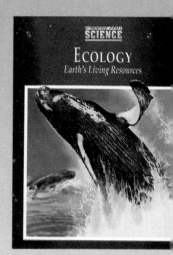

Fit for All Your Curriculum Needs.

The flexibility of this program will allow you to teach only those topics you want to teach, and to teach them *in-depth*. Virtually any approach to science—general, integrated, coordinated, thematic, etc.—is possible with *Prentice Hall Science*.

Above all, the program is designed to make your teaching experience easier and more fun.

For more detailed information about each title, see pages 16–34.

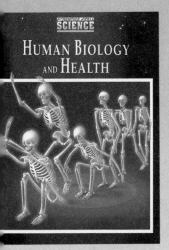

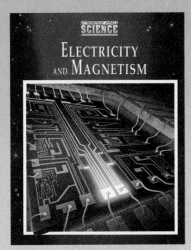

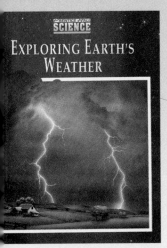

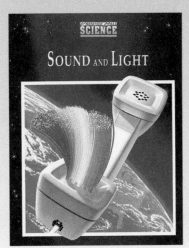

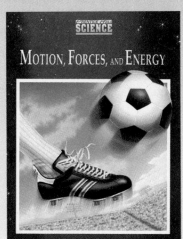

A Completely Integrated Learning System . .

The *Prentice Hall Science* program is an *integrated* learning system with a variety of print materials and multimedia components. All are designed to meet the needs of diverse learning styles and your technology needs.

☐ Student Edition
☐ Teacher's Resource Package

In addition to the Student Edition, the Teacher's Resource Package includes:

- Annotated Teacher's Edition
- Spanish Audiotape
- English Audiotape
- Activity Book
- Review and Reinforcement Guide
- Test Book
- Laboratory Manual, Annotated Teacher's Edition

For the Perfect Fit to Your Teaching Needs.

❏ Teacher's Desk Reference
❏ Product Testing Activities
❏ Laboratory Manual, Student Edition
❏ English Guide for Language Learners
❏ Spanish Guide for Language Learners
❏ Transparencies
❏ Computer Test Bank Disks (IBM, Apple, or MAC)
❏ Videodiscs
❏ Interactive Videodiscs (Level III)
❏ Interactive Videodiscs/CD ROM

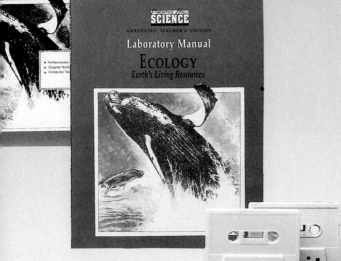

All components are integrated in the teaching strategies in the Annotated Teacher's Edition, where they directly relate to the science content.

For more detailed information about each component, see pages 10–15 and 35–39.

The Test Book in each Teacher's Resource Package offers many assessment alternatives for each chapter.

Performance-Based Tests focus on science processes and skills and are designed for students to perform at individual workstations. These authentic assessment situations require students to demonstrate their problem-solving and thinking skills using real objects and phenomena.

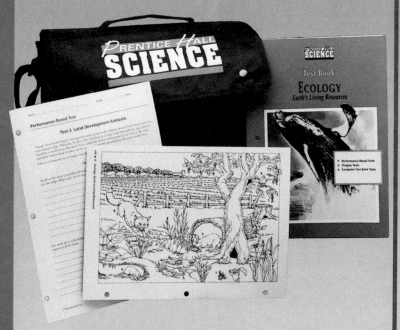

Computer Test Bank questions may be used to generate your own style of tests. Just insert the disk into your Apple, MAC, or IBM pc and choose the multiple choice, true/false, completion, and critical-thinking questions that meet the assessment needs of your individual classes—from review sheets to quizzes, chapter tests, and finals.

Chapter tests include the same types of questions as the **Computer Test Bank** but may be used as is for testing purposes or review sheets.

The **Annotated Teacher's Edition** includes other assessment alternatives, including suggestions for Portfolio Assessment.

Prentice Hall Science offers curriculum-based Interactive Videodisc programs, plus Interactive Videodisc/CD ROM programs, designed for the ultimate integration of concepts, problem-solving skills, and relevancy. As part of the Learning System, the programs are integrated within the teaching strategies of the Annotated Teacher's Edition.

For interactive videodiscs, all you need is a videodisc player and a television monitor controlled by a computer. Add a CD ROM drive and you are ready to use interactive videodisc/CD ROM programs as well. These technologies allow students to interact with captivating science programs. Intriguing problems and activities stimulate students' creativity as they become involved in this new approach to learning.

Interactive technologies have revolutionized student independent research and multimedia reports. They're also great for classroom projects and group learning activities. And the vast information supplied by interactive technologies offers more flexibility for your teaching needs and more discovery opportunities for your students.

The Discovery Interactive Library™ is a series of documentary-based interactive video programs from The Discovery Channel®. Programs include

Insects: Little Giants of the Earth
Invention: Mastering Sound
Planet Earth: The Force Within
Planet Earth: The Blue Planet
Planet Earth: The Solar Sea
In the Company of Whales
On Dry Land: The Desert Biome
Investigating Science: Treasures From the Deep

You have the option of purchasing any program in the Discovery Interactive Library in videodisc format only for Level I use. With this package you receive one two-sided videodisc and a Classroom Guide including an index, barcodes, transcript, keyword index, glossary, and more. You may also choose to purchase computer software in combination with the Level I package for Level III interactive use.

The ScienceVision™ **Series** is a series of six inquiry-based videodisc packages funded by the National Science Foundation and developed in conjunction with the Interactive Media Science Department at Florida State University and Houghton Mifflin. Titles include

Science Vision: ErgoMotion
Science Vision: EcoVision
Science Vision: AstroVision
Science Vision: BioExplorer
Science Vision: Chemical Pursuits
Science Vision: TerraVision

With each ScienceVision program, you get one two-sided videodisc, a Teacher's Guide, a User's Guide, two Activity Books, and computer software.

The Interactive Videodisc/CD ROM Library contains four programs that include videodiscs plus CD ROM discs to provide you with the latest advances in educational technology. Each program comes complete with Teacher's Guides and all the instructions to get you up and running. The four interactive videodisc/CD ROM programs are:

Virtual BioPark and Amazonia, jointly developed by Prentice Hall and the Smithsonian Institute, enable students to explore living things, concentrating on the interrelationships between organisms and their environments.

Paul ParkRanger and the Mystery of the Disappearing Ducks, developed under the auspices of Lucas-Film, provides students with a real life mystery— what has happened to the ducks?

Science Discovery, developed by Computer Curriculum Corporation, provides students with video and simulations on 32 topics related to life, earth, and physical science.

To find out the programs integrated within each *Prentice Hall Science* book, see pages 16–34.

To learn more about integrating technology into your classroom, see pages 78–79.

*P*rentice Hall Science offers you the opportunity to make exciting **video-discs** part of your teaching strategy.

The teaching strategies in the wraparound margins of the *Annotated Teacher's Edition* tell you exactly when it would be appropriate to show a specific videodisc. Included in the strategies are suggested follow-up activities. In addition, each videodisc includes a Discussion Guide and further activity suggestions for more discovery learning.

There are 32 titles from which to choose:
Super Scents
Seeing Sense
Sound Sense
Patterns of Evolution
Biomes: Deserts and Tundra
Biomes: Coniferous Forest and Tropical Rain Forest
Biomes: Temperate Deciduous Forest and Grassland
Aquatic Ecosystems: Freshwater Wetlands and Freshwater
Aquatic Ecosystems: Estuaries and Marine
Earth's Atmosphere

Wind and Air Currents
Atmospheric Series: Atmosphere and Winds and Air Currents
Atmospheric Series: Clouds and Precipitation and Global Winds
Atmospheric Series: Violent Storms and Weather Systems in Motion
Elements, Compounds, and Mixtures
Periodic Table and Periodicity
Chemical Bonding and Atomic Structure
Acids, Bases, and Salts
Systems Working Together
Muscular System
Skeletal System
Digestive System
Excretory System
Respiratory System
Endocrine System
Nervous System
Circulatory System
Reproductive System

See pages 16–34 to find out which titles are integrated into each of the 19 *Prentice Hall Science* books.

Activity Bank — New for Second Edition

A special feature called the Activity Bank ends each textbook in *Prentice Hall Science*. The Activity Bank is a compilation of hands-on activities designed to develop and reinforce science concepts presented in each chapter of the textbook. Activity Bank activities provide opportunities to meet the diverse abilities and interests of students; to encourage problem solving, critical thinking, and discovery learning; to involve students more actively in the learning experience; and to address the need for ESL strategies and Cooperative Learning.

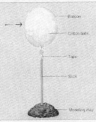

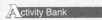

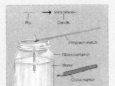

Whenever you see this logo in the margins of the textbook, you will know that an additional hands-on activity that further develops the concepts of that section can be found at the back of the textbook in the Activity Bank section. All Activity Bank activities are reproduced in a blackline master format in the back of the Activity Book. In addition, even more hands-on activities have been included in the blackline master Activity Bank at the end of the Activity Book, providing you with a wealth of extra hands-on opportunities.

Integrated Science Activity Book For those teachers who want to fully integrate activities from diverse areas of science into their course of study, Prentice Hall has developed the Integrated Science Activity Book. This book contains all of the blackline master activities found at the back of the 19 Activity Books that make up *Prentice Hall Science*. For convenience of use, these activities are grouped into Life Science, Earth Science, and Physical Science activities. Using the Integrated Science Activity Book, you can now incorporate related activities from various fields of science into the science topics you are currently teaching.

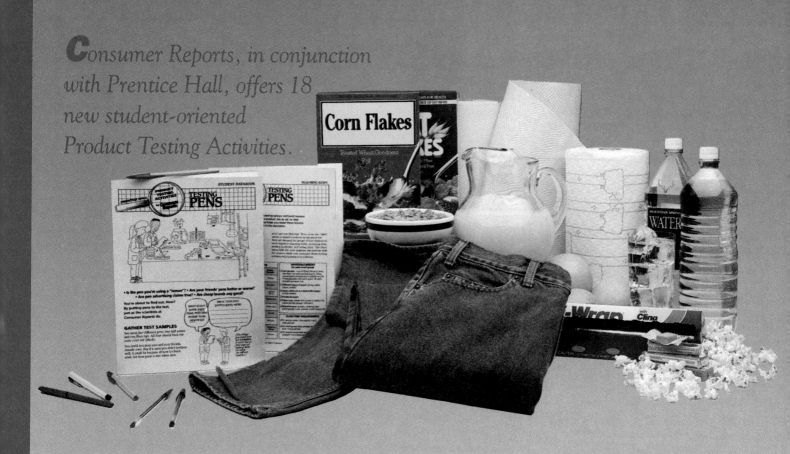

Consumer Reports, in conjunction with Prentice Hall, offers 18 new student-oriented Product Testing Activities.

Testing Antacids

Testing Bandages

Testing Bottled Water

Testing Bubble Gum

Testing Cereals

Testing Disposable Cups

Testing Food Wraps

Testing Glues

Testing Jeans

Testing Lip Balms

Testing Nail Enamels

Testing Orange Juice

Testing Paper Towels

Testing Popcorn

Testing Shampoos

Testing Sports Drinks

Testing Toilet Paper

Testing Yogurts

Students test these products based on the same methodologies used by the science teams at Consumer Reports, then devise their own methodologies for testing additional variables. Students experience first-hand the science concepts related to each product and understand how these concepts relate to their lives.

The kits are hands-on, motivational, and modeled on product testing done by *Consumer Reports Magazine*. Because the activities test everyday items, they do not require any extensive equipment. They're easy and fun to teach, and, most importantly, they address the following student needs:

- Problem-solving

- Applications to everyday life

- Motivation/creating an interest in science

- Science, technology, and society

- Hands-on, process science

- Thinking and understanding through involvement

- Cooperative learning

- Parent involvement

Transparencies

Prentice Hall Science offers **70 full-color transparencies** that illustrate key concepts from each of the 19 *Prentice Hall Science* books. You get them all, for times when you would like to visually make a connection to another science field of study. The transparency package includes 36 pages of teacher support, including questioning strategies that promote critical thinking and concept mastery.

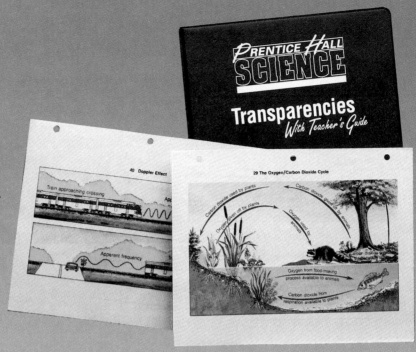

Materials for Students of Limited English Ability

Prentice Hall Science offers two related components designed to teach students of limited English ability. The *Spanish Guide for Language Learners Audiotapes* and *English Guide for Language Learners Audiotapes*, with separate print versions available, can be used with Limited English Proficiency (LEP) students or any student who may need special help in developing English language skills.

The audiotapes walk students through each section of the textbook, focusing their attention on key ideas and terms, as well as on photographs and artwork that will help them visually grasp the meaning of the concepts. The review questions at the end of each section in the text are also narrated on the tape. Both the English and Spanish audiotapes are packaged in the individual Teacher's Resource Packages for each *Prentice Hall Science* title. The English and Spanish print versions, which contain scripts for all 19 books, are available separately and offer translation rights into any language. The print version of the *Spanish Guide for Language Learners* includes a Spanish glossary.

The Nature of Science

Ch.1 What Is Science?

Ch.2 Measurement and the Sciences

Ch.3 Tools and the Sciences

Primary Themes Scale and Structure, Systems and Interactions, and Unity and Diversity.

THE NATURE OF SCIENCE develops the attitudes and skills necessary for the successful study of science. Students become scientists as they observe and question events in their world. As they are taken step-by-step through the scientific method of problem-solving, students begin to understand that in practice these "steps" may not occur in such orderly fashion. Indeed, as the text discusses, serendipity can be an important factor in scientific discovery.

Before setting out to explore the world of science, including its methods and tools, students must be made aware of the safety rules involved—as they are in Chapter 1 in this book and the 18 other books in *Prentice Hall Science*. From here they are taught the metric system and methods of measurement as well as an in-depth introduction to the concepts of mass, volume, and density.

Finally, students become familiar with the basic tools of the lab through illustrations, discussion, and activities. In addition, the contributions of advanced technology in scientific equipment are explored.

The other *Prentice Hall Science* titles integrated within THE NATURE OF SCIENCE are

Evolution: Change Over Time
Parade of Life: Animals
Exploring the Universe
Heat Energy
Parade of Life: Monerans, Protists, Fungi, and Plants
Dynamic Earth
Ecology: Earth's Natural Resources
Motion, Forces, and Energy
Matter: Building Block of the Universe
Sound and Light
Exploring Earth's Weather

MEDIA AND TECHNOLOGY

In addition to the Student Edition, Annotated Teacher's Edition, Teacher's Resource Package, and Laboratory Manual, the following media and technology components are integrated within the teaching strategies for THE NATURE OF SCIENCE:

Interactive Videodisc/CD ROM

Paul ParkRanger and the Mystery of the Disappearing Ducks

Product-Testing Activities

Testing Bubble Gum
Testing Shampoo
Testing Cereal
Testing Pens

Parade of Life: Monerans, Protists, Fungi, and Plants

Ch.1 Classification of Living Things

Ch.2 Viruses and Monerans

Ch.3 Protists

Ch.4 Fungi

Ch.5 Plants Without Seeds

Ch.6 Plants With Seeds

Primary Themes Evolution, Patterns of Change, Scale and Structure, and Unity and Diversity.

Any parade would prove chaotic without a theme to hold it together, and the parade of life is no different. Therefore, this book begins with a discussion of the five-kingdom biological system of classification and its evolutionary basis.

The theme of evolution is carried throughout the remaining chapters, providing a basis for in-depth discussion of viruses and monerans, protists, fungi, and plants. Students learn the characteristics of members of each of these kingdoms, as well as the importance of such organisms in our world.

The discussion of plants is divided into two parts. The study of plants without seeds includes algae, mosses, and ferns. In investigating plants with seeds, students discover the characteristics and functions of roots, stems, and leaves, and the role and mechanics of photosynthesis.

The other *Prentice Hall Science* titles integrated within PARADE OF LIFE: MONERANS, PROTISTS, FUNGI, AND PLANTS are

Evolution: Change Over Time
Human Biology and Health
Cells: Building Blocks of Life
Ecology: Earth's Living Resources
Exploring Planet Earth
Dynamic Earth
Matter: Building Block of the Universe
Heat Energy
Exploring the Universe
Chemistry of Matter
Ecology: Earth's Natural Resources

In addition to the Student Edition, Annotated Teacher's Edition, Teacher's Resource Package, and Laboratory Manual, the following media and technology components are integrated within the teaching strategies for PARADE OF LIFE: MONERANS, PROTISTS, FUNGI, AND PLANTS:

MEDIA AND TECHNOLOGY

Interactive Videodiscs/CD ROM

Paul ParkRanger and the Mystery of the Disappearing Ducks
Virtual BioPark
Amazonia

Interactive Videodiscs

Science Vision: EcoVision
On Dry Land: The Desert Biome

Product-Testing Activities

Testing Yogurt
Testing Jeans

Transparencies

The Bacteriophage Virus
Protists
Mushroom Development
Bread Mold

Parade of Life: Animals

Ch.1 Sponges, Cnidarians, Worms, and Mollusks

Ch.2 Arthropods and Echinoderms

Ch.3 Fishes and Amphibians

Ch.4 Reptiles and Birds

Ch.5 Mammals

Primary Themes Evolution, Patterns of Change, Scale and Structure, and Unity and Diversity.

In this text, students venture into the wonderful world of animals, discovering the characteristics, roles, and evolutionary relationships of the various phyla. Through exciting prose, visuals, and activities, students are introduced to a cast of characters including sponges, corals, worms, and mollusks. The exploration continues in Chapter 2 as the text reveals the mysteries of starfish, crustaceans, and those amazing insects.

The book continues with the evolution of vertebrates, starting with the fishes. Moving from sea to land, students learn about the adaptations made by amphibians for their dual lifestyles, eventually leading to land animals—reptiles, birds, and mammals.

The other *Prentice Hall Science* titles integrated within PARADE OF LIFE: ANIMALS are

Parade of Life: Monerans, Protists, Fungi, and Plants
Ecology: Earth's Living Resources
Cells: Building Blocks of Life
Parade of Life: Animals
Evolution: Change Over Time
Exploring Planet Earth
Human Biology and Health
Sound and Light
Electricity and Magnetism

MEDIA AND TECHNOLOGY

In addition to the Student Edition, Annotated Teacher's Edition, Teacher's Resource Package, and Laboratory Manual, the following media and technology components are integrated within the teaching strategies for PARADE OF LIFE: ANIMALS:

Interactive Videodiscs/CD ROM

Paul ParkRanger and the Mystery of the Disappearing Ducks
Amazonia
Virtual BioPark

Interactive Videodiscs

Science Vision: EcoVision
Insects: Little Giants of the Earth
On Dry Land: The Desert Biome
In the Company of Whales

Videos/Videodiscs

Super Scents
Seeing Sense
Sound Sense

Cells: Building Blocks of Life

Ch.1 The Nature of Life

Ch.2 Cell Structure and Function

Ch.3 Cell Processes

Ch.4 Cell Energy

Primary Themes Energy, Evolution, Scale and Structure, Systems and Interactions, and Stability.

Students begin their study of the cell by first learning about the origin of life on Earth. To help them understand this concept, the program explores the early Earth, the formation of the molecules of life, and how cells formed and evolved into modern cells. Students then discover the characteristics shared by all living things, as well as their common needs. Concluding the chapter is a study about the chemistry of living things.

Chapter 2 begins with an introduction to the cell theory. Then it's off on an imaginary journey through a living organism, visiting its tiny cells and observing some of the typical structures and their roles in the functioning of the cell. Students discover each role's relationship to cell processes such as diffusion and osmosis, cell growth, and division. The book concludes with a look at cell energy, including coverage of photosynthesis and respiration.

The other *Prentice Hall Science* titles integrated within CELLS: BUILDING BLOCKS OF LIFE are

Exploring the Universe
Evolution: Change Over Time
Exploring Planet Earth
Parade of Life: Monerans, Protists, Fungi, and Plants
Parade of Life: Animals
Ecology: Earth's Living Resources
Chemistry of Matter
Dynamic Earth
Human Biology and Health
Heredity: The Code of Life
Ecology: Earth's Natural Resources

In addition to the Student Edition, Annotated Teacher's Edition, Teacher's Resource Package, and Laboratory Manual, the following media and technology components are integrated within the teaching strategies for CELLS: BUILDING BLOCKS OF LIFE:

MEDIA AND TECHNOLOGY

Interactive Videodisc/CD ROM
Science Discovery: Cells

Product-Testing Activities
Testing Bottled Water
Testing Sports Drinks
Testing Paper Towels

Transparencies
Redi's Experiment
Cell Functions
Typical Plant and Animal Cells
The Oxygen/Carbon Dioxide Cycle

Heredity: The Code of Life

Ch.1 What Is Genetics?   Ch.3 Human Genetics

Ch.2 How Chromosomes Work   Ch.4 Applied Genetics

Primary Themes Evolution, Patterns of Change, Systems and Interactions, and Unity and Diversity.

This book traces the history of genetics to its modern understanding of genes and chromosomes. Students gain a basic understanding of genetics in Chapter 1 as they explore the principles of heredity through the work of Gregor Mendel and Karl Correns. Students predict the results of genetic crosses by using their new-found knowledge of probability and Punnett Squares.

In discovering how chromosomes work, students first learn about the chromosome theory, the primary functions of chromosomes in heredity, and mutations. They go on to learn about the DNA molecule and its role in heredity.

The basic principles of heredity are applied to human genetics in Chapter 3. Included in this chapter is a study of sex-linked traits and human genetic disorders. In the final chapter, students discover the many applications of genetics and its impact on society.

The other *Prentice Hall Science* titles integrated within HEREDITY: THE CODE OF LIFE are

Parade of Life: Monerans, Protists, Fungi, and Plants
Parade of Life: Animals
Cells: Building Blocks of Life
Human Biology and Health
Ecology: Earth's Living Resources

MEDIA AND TECHNOLOGY In addition to the Student Edition, Annotated Teacher's Edition, Teacher's Resource Package, and Laboratory Manual, the following media and technology components are integrated within the teaching strategies for HEREDITY: THE CODE OF LIFE:

Interactive Videodisc/CD ROM
Virtual BioPark

Interactive Videodisc
Science Vision: BioExplorer

Transparencies
Mendel's Peas
Seven Characteristics of Pea Plants
Independent Assortment
Incomplete Dominance
Punnett Squares
Recombinant DNA

Evolution: Change Over Time

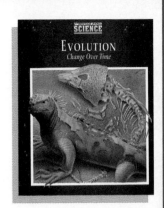

Ch.1 Earth's History in Fossils

Ch.2 Changes in Living Things Over Time

Ch.3 The Path to Modern Humans

Primary Themes Evolution, Patterns of Change, Systems and Interactions, and Unity and Diversity.

Students begin their investigation of Earth's history by discovering how scientists use fossils as clues to the past. The answers to "What is a fossil?" and "How do fossils form?" are found in the first chapter, as well as techniques for dating fossils. Students are introduced to relevant topics in geology, such as faults, intrusions, and extrusions. The chapter concludes with a trip through time to explore the major geologic eras.

In Chapter 2, students learn how the recording of Earth's history through fossils clearly demonstrates how living things evolved over time. Anatomical, embryological, chemical, and molecular evidence of evolution is also examined. The chapter outlines the history of natural selection and relates the concept to overproduction and variation among species. This leads into how a new species evolves, the role of adaptive radiation, and punctuated equilibrium.

Students join the search for human ancestors in the final chapter as they trace the evolutionary path to modern humans.

The other *Prentice Hall Science* titles integrated within EVOLUTION: CHANGE OVER TIME are

Dynamic Earth
Exploring Planet Earth
Exploring Earth's Weather
Chemistry of Matter
Exploring the Universe
Cells: Building Blocks of Life
Parade of Life: Monerans, Protists, Fungi,
   and Plants
Heredity: The Code of Life
Parade of Life: Animals
Ecology: Earth's Living Resources
Human Biology and Health
Sound and Light

In addition to the Student Edition, Annotated Teacher's Edition, Teacher's Resource Package, and Laboratory Manual, the following media and technology components are integrated within the teaching strategies for EVOLUTION: CHANGE OVER TIME:

MEDIA AND TECHNOLOGY

Interactive Videodiscs/CD ROM

Paul ParkRanger and the Mystery of the
  Disappearing Ducks
Virtual BioPark
Amazonia
Science Discovery: Plate Tectonics

Interactive Videodiscs

Science Vision: TerraVision
Insects: Little Giants of the Earth

Video/Videodisc

Patterns of Evolution

Transparencies

Half–Life
Fossil Links

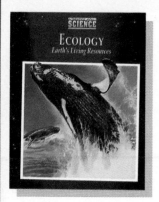

Ecology: Earth's Living Resources

Ch. 1 Interactions Among Living Things

Ch. 2 Cycles in Nature

Ch. 3 Exploring Earth's Biomes

Ch. 4 Wildlife Conservation

Primary Themes Energy, Evolution, Patterns of Change, Systems and Interactions, and Unity and Diversity.

Students are introduced to their study of ecology with an important concept—that all living and nonliving things in an environment are interconnected. Throughout Chapter 1, students discover various relationships and how a disturbance in one part of an ecosystem can affect the entire ecosystem. Chapter 2 explores the various cycles in nature and how they affect organisms and ecosystems. Students discover cycles in time, matter, and change.

By taking a journey through six of Earth's major biomes, students explore in Chapter 3 the relationships among plants, animals, and their environment. The final chapter focuses on the importance—and the problems—of wildlife conservation. Students increase their awareness of their role in preserving Earth's living resources as they realize that they are also interconnected with the environment. The book concludes with a search for solutions and a philosophy that can help students live with respect for the environment.

The other *Prentice Hall Science* titles integrated within ECOLOGY: EARTH'S LIVING RESOURCES are

Exploring Earth's Weather
Exploring the Universe
Dynamic Earth
Evolution: Change Over Time
Exploring Planet Earth
Parade of Life: Monerans, Protists, Fungi, and Plants
Human Biology and Health

MEDIA AND TECHNOLOGY

In addition to the Student Edition, Annotated Teacher's Edition, Teacher's Resource Package, and Laboratory Manual, the following media and technology components are integrated within the teaching strategies for ECOLOGY: EARTH'S LIVING RESOURCES:

Interactive Videodiscs/CD ROM
Paul ParkRanger and the Mystery of the Disappearing Ducks
Virtual BioPark
Amazonia

Interactive Videodiscs
Science Vision: EcoVision
In the Company of Whales
Insects: Little Giants of the Earth
On Dry Land: The Desert Biome

Videodiscs
Biomes: Desert and Tundra
Biomes: Coniferous Forest and Tropical Rain Forest
Biomes: Temperate Deciduous Forest and Grassland

Aquatic Ecosystems: Freshwater Wetlands and Freshwater
Aquatic Ecosystems: Estuaries and Marine

Product-Testing Activities
Testing Toilet Paper
Testing Orange Juice
Testing Disposable Cups
Testing Food Wraps

Transparencies
Food Web
Energy Pyramid
Nitrogen Cycle
Oxygen/Carbon Dioxide Cycle

Human Biology and Health

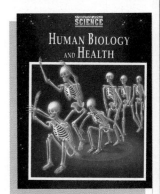

Primary Themes Energy, Scale and Structure, Systems and Interactions, and Stability.

Students begin their exploration of the incredible human machine by discovering how the body's activities work together to maintain a stable internal environment. After they gain an understanding of the levels of organization in living organisms, they move on to the individual systems.

The book covers the structure and function of the ten basic systems of the body. Integrated throughout is appropriate information on health and safety. As students explore the immune system, they learn about the AIDS virus and its impact on society.

The other *Prentice Hall Science* titles integrated within HUMAN BIOLOGY AND HEALTH are

Cells: Building Blocks of Life
Chemistry of Matter
Motion, Forces, and Energy
Parade of Life: Animals
Dynamic Earth
Electricity and Magnetism
Heat Energy
Ecology: Earth's Natural Resources
Exploring Planet Earth
Sound and Light
Exploring the Universe
Parade of Life: Monerans, Protists, Fungi, and Plants
Ecology: Earth's Living Resources
Heredity: The Code of Life

In addition to the Student Edition, Annotated Teacher's Edition, Teacher's Resource Package, and Laboratory Manual, the following media and technology components are integrated within the teaching strategies for HUMAN BIOLOGY AND HEALTH:

MEDIA AND TECHNOLOGY

Interactive Videodisc
Science Vision: BioExplorer

Videos/Videodiscs
Systems Working Together
Muscular System
Skeletal System
Digestive System
Excretory System
Respiratory System
Endocrine System
Nervous System
Circulatory System
Reproductive System

Product-Testing Activities
Testing Bandages
Testing Cereals
Testing Orange Juice
Testing Sports Drinks
Testing Yogurts
Testing Antacids

Transparencies
Joints
Circulatory System
The Heart
Respiratory System
Urinary System
Three Types of Neurons
The Skin
The Eye
The Ear
Male Reproductive System
Female Reproductive System

Exploring Planet Earth

Ch.1 Earth's Atmosphere   Ch.4 Earth's Landmasses
Ch.2 Earth's Oceans   Ch.5 Earth's Interior
Ch.3 Earth's Fresh Water

Primary Themes   Energy, Patterns of Change, Scale and Structure, and Systems and Interactions.

The exploration of Earth begins with an overall view of the planet. From here students take an in-depth look at the Earth's early atmosphere and how it has evolved over time. The Earth's present atmosphere is studied in terms of its make-up, its various layers, and the effects of pollution.

Next, students examine the properties of ocean water and discover the features of the ocean floor. The adventure continues as they explore the amazing creatures in each of the ocean's major life zones. The chapter concludes with the motions of the oceans.

In Chapter 3, students recognize the importance of the Earth's fresh water and discover its sources. The role of water as a solvent is examined through a study of the composition of water and some of its properties.

In Chapter 4, students study the Earth's continents and its topography as they practice their mapping skills. Finally, students are guided through the Earth's interior, discovering the characteristics and phenomena associated with the core, mantle, and crust.

The other *Prentice Hall Science* titles integrated within EXPLORING PLANET EARTH are

Exploring the Universe
Chemistry of Matter
Sound and Light
Cells: Building Blocks of Life
Human Biology and Health
Ecology: Earth's Living Resources
Exploring Earth's Weather
Dynamic Earth
Heat Energy
Matter: Building Block of the Universe
Parade of Life: Monerans, Protists, Fungi, and Plants
Ecology: Earth's Natural Resources
Parade of Life: Animals
Electricity and Magnetism

MEDIA AND TECHNOLOGY

In addition to the Student Edition, Annotated Teacher's Edition, Teacher's Resource Package, and Laboratory Manual, the following media and technology components are integrated within the teaching strategies for EXPLORING PLANET EARTH:

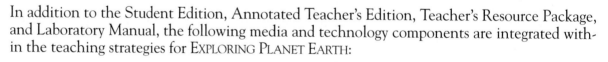

Interactive Videodisc/CD ROM
Amazonia

Interactive Videodiscs
Science Vision: TerraVision
Science Vision: EcoVision
Planet Earth: The Force Within
Planet Earth: The Blue Planet
On Dry Land: The Desert Biome
Investigating Science: Treasures From the Deep

Videodiscs
Aquatic Ecosystems: Freshwater Wetlands and Freshwater
Aquatic Ecosystems: Estuaries and Marine

Videos/Videodiscs
Earth's Atmosphere
Wind and Air Currents

Product-Testing Activity
Testing Bottled Water

Transparencies
Ocean-Floor Features
Ocean Wave Characteristics
Long-Distance Surface Currents
Rip Currents
The Water Cycle
S and P Waves
Landscape Regions

Dynamic Earth

Ch.1 Movement of the Earth's Crust   Ch.4 Rocks and Minerals

Ch.2 Earthquakes and Volcanoes   Ch.5 Weathering and Soil Formation

Ch.3 Plate Tectonics   Ch.6 Erosion and Deposition

Primary Themes   Energy, Evolution, Patterns of Change, Scale and Structure, and Stability.

In this book, students discover the magnificent forces that cause Earth to be in a constant state of change. The first chapter delves into deformities caused by stress, the major types of faulting and folding, how domes and plateaus are formed, and the effects of isostasy on the Earth's crust. The exploration continues as students explore the phenomena associated with earthquakes and volcanoes and identify the major zones of activity.

In Chapter 3, students explore the theory of continental drift, ocean-floor spreading, and the theory of plate tectonics. Fossil and rock evidence is examined and related to continental drift.

To understand what minerals are, students examine their characteristics and how they are formed. They are then given plenty of opportunities to use their knowledge by identifying various minerals. The same steps are followed to understand rocks and the rock cycle as students take an in-depth look at igneous, sedimentary, and metamorphic rocks.

In studying weathering and soil formation, students discover how mechanical and chemical weathering affect the surface of the Earth. They investigate soil formation and composition, as well as the factors that af-

fect soil formation. Finally, students examine the types of erosion and deposition and identify the various agents which cause them.

The other *Prentice Hall Science* titles integrated within DYNAMIC EARTH are

> Exploring Planet Earth
> Matter: Building Block of the Universe
> The Nature of Science
> Evolution: Change Over Time
> Parade of Life: Animals
> Exploring Earth's Weather
> Electricity and Magnetism
> Heat Energy
> Chemistry of Matter
> Ecology: Earth's Natural Resources
> Ecology: Earth's Living Resources
> Parade of Life: Monerans, Protists, Fungi, and Plants
> Dynamic Earth
> Motion, Forces, and Energy

In addition to the Student Edition, Annotated Teacher's Edition, Teacher's Resource Package, and Laboratory Manual, the following media and technology components are integrated within the teaching strategies for DYNAMIC EARTH:

MEDIA AND TECHNOLOGY

Interactive Videodiscs/CD ROM

Amazonia
Science Discovery: Plate Tectonics

Interactive Videodiscs

Science Vision: TerraVision
Planet Earth: The Force Within
Planet Earth: The Blue Planet
On Dry Land: The Desert Biome
Investigating Science: Treasures From the Deep

Video/Videodisc

Earth's Atmosphere

Product-Testing Activity

Testing Nail Enamels

Transparencies

Fossil Links
S and P Waves
Lithospheric Plates
Soil Profile
Ocean Wave Characteristics
Long-Distance Surface Currents
Rip Currents

Exploring Earth's Weather

Ch. 1 What Is Weather?

Ch. 2 What Is Climate?

Ch. 3 Climate in the United States

Primary Themes Energy, Patterns of Change, and Systems and Interactions.

In order to answer the question "What is weather?" students investigate and discover the factors that cause weather. As each of the factors are examined in depth, students become aware of the relationships among them and how they are measured. In studying moisture in the air, students also learn to identify basic cloud types. After they understand the factors affecting weather, students go on to investigate weather patterns and gain experience in observing and predicting weather.

In Chapter 2, students discover the answers to "What is climate and what causes it?" and "How do temperature and precipitation influence climate?" The Earth's major climate zones are explored, and the four seasons are related to the tilt of the Earth's axis. In studying the evolution of Earth's climate, students become aware of how changes in climate affect living things. They also examine the issues surrounding the extinction of dinosaurs, as well as global warming.

Finally, students explore the kinds of climates found in the United States and classify its land biomes.

The other *Prentice Hall Science* titles integrated within EXPLORING EARTH'S WEATHER are

Exploring Planet Earth
Parade of Life: Animals
Exploring the Universe
Heat Energy
Ecology: Earth's Natural Resources
Matter: Building Block of the Universe
Electricity and Magnetism
Sound and Light
Ecology: Earth's Living Resources
Dynamic Earth
Evolution: Change Over Time
Heredity: The Code of Life
Parade of Life: Monerans, Protists, Fungi, and Plants
Human Biology and Health

MEDIA AND TECHNOLOGY

In addition to the Student Edition, Annotated Teacher's Edition, Teacher's Resource Package, and Laboratory Manual, the following media and technology components are integrated within the teaching strategies for EXPLORING EARTH'S WEATHER:

Interactive Videodisc/CD ROM
Amazonia

Interactive Videodiscs
Planet Earth: The Blue Planet
On Dry Land: The Desert Biome
Insects: Little Giants of the Earth

Videodiscs
Biomes: Desert and Tundra
Biomes: Coniferous Forest and Tropical Rain Forest
Biomes: Temperate Deciduous Forest and Grassland
Atmospheric Series: Atmosphere and Winds and Air Currents

Atmospheric Series: Clouds and Precipitation and Global Winds
Atmospheric Series: Violent Storms and Weather Systems in Motion
Aquatic Ecosystems: Freshwater Wetlands and Freshwater
Aquatic Ecosystems: Estuaries and Marine

Product-Testing Activities
Testing Bandages
Testing Disposable Cups
Testing Food Wraps

Transparencies
The Water Cycle
Long-Distance Surface Currents

Ecology: Earth's Natural Resources

Ch.1 Energy Resources

Ch.3 Pollution

Ch.2 Earth's Nonliving Resources

Ch.4 Conserving Earth's Resources

Primary Themes Energy, Scale and Structure, Systems and Interactions, and Stability.

Students discover the energy options for today and tomorrow while learning about combustion and fossil fuels, their uses, and their conservation. Energy from the sun, wind, water, and nuclear fission and fusion are analyzed in terms of how they work, their advantages and disadvantages, and their promises for the future. Other alternative energy resources included are geothermal energy, tidal energy, biomass, and hydrogen power.

Students increase their awareness of Earth's assets as they identify how people make use of land and soil resources, fresh water, and minerals. They discover how Earth is like a giant spaceship with a living cargo but carrying a limited amount of vital supplies. This sets the tone for an in-depth and active look at pollution in Chapter 3, which ends with a discussion on what can be done about the problems we face.

The last chapter takes an active look at how people, including junior citizens, can make a difference with the wise use of natural resources. Various methods of conserving fossil fuels are investigated, as well as means of protecting the environment.

The other *Prentice Hall Science* titles integrated within ECOLOGY: EARTH'S NATURAL RESOURCES are

Dynamic Earth
Heat Energy
Chemistry of Matter
Electricity and Magnetism
Sound and Light
Exploring Planet Earth
Exploring the Universe
Ecology: Earth's Living Resources
Exploring Earth's Weather
Parade of Life: Monerans, Protists, Fungi, and Plants

In addition to the Student Edition, Annotated Teacher's Edition, Teacher's Resource Package, and Laboratory Manual, the following media and technology components are integrated within the teaching strategies for ECOLOGY: EARTH'S NATURAL RESOURCES:

MEDIA AND TECHNOLOGY

Interactive Videodisc/CD ROM
Paul ParkRanger and the Mystery of the Disappearing Ducks

Interactive Videodiscs
Science Vision: TerraVision
Science Vision: EcoVision
Science Vision: Chemical Pursuits
Planet Earth: The Blue Planet

Video/Videodisc
Earth's Atmosphere

Product-Testing Activities
Testing Disposable Cups
Testing Food Wraps
Testing Paper Towels

Transparencies
Solar Heating System
Chain Reaction

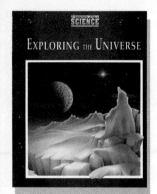

Exploring the Universe

Ch.1 Stars and Galaxies

Ch.2 The Solar System

Ch.3 Earth and Its Moon

Primary Themes Energy, Evolution, Scale and Structure, Systems and Interactions, and Unity and Diversity.

Buckle your seatbelts and prepare for a trip through the universe! Along the way, students discover multiple star systems, constellations, novas, nebulas, and galaxies. They also investigate how this extraordinary universe formed, the Big Bang Theory, and some of the tools and techniques used by astronomers to study distant stars. The characteristics and evolution of stars are studied, including a very special star—the sun.

Next, students investigate the evolution of the solar system, starting with the nebular theory and comparing the formation of the inner and outer planets. From there, they study and analyze the motions of the planets—the shape of their orbits and periods of revolution and rotation. Armed with this knowledge, students once again board their imaginary spaceship and take off to observe the characteristics of the planets and other objects in the solar system. Real-life adventures are discovered as students learn how a reaction engine works and the contributions of the various spacecraft sent to probe the solar system.

The adventure isn't over yet, as students explore the Earth's place in the solar system and the relationship between the Earth and its moon. Students investigate the characteristics of the moon, several theories for the origin of the moon, and identify the interactions among the Earth, the moon, and the sun. The chapter wraps up with a look at the importance of space technology to society.

The other *Prentice Hall Science* titles integrated within EXPLORING THE UNIVERSE are

Sound and Light
The Nature of Science
Motion, Forces, and Energy
Chemistry of Matter
Heat Energy
Ecology: Earth's Natural Resources
Matter: Building Block of the Universe
Evolution: Change Over Time
Dynamic Earth
Exploring Earth's Weather
Electricity and Magnetism
Cells: Building Blocks of Life
Exploring Planet Earth

MEDIA AND TECHNOLOGY In addition to the Student Edition, Annotated Teacher's Edition, Teacher's Resource Package, and Laboratory Manual, the following media and technology components are integrated within the teaching strategies for EXPLORING THE UNIVERSE:

Interactive Videodiscs
Science Vision: AstroVision
The Solar Sea

Product-Testing Activities
Testing Sports Drinks
Testing Pens

Transparencies
Gravitational Pull
Escape Velocity
Hertzsprung-Russell Diagram
Layers of the Sun
Life Cycle of a Star
Doppler Effect
Electromagnetic Spectrum

Matter: Building Block of the Universe

Ch.1 General Properties of Matter

Ch.2 Physical and Chemical Changes

Ch.3 Mixtures, Elements, and Compounds

Ch.4 Atoms: Building Blocks of Matter

Ch.5 Classification of Elements: The Periodic Table

Primary Themes Patterns of Change, Scale and Structure, Systems and Interactions, and Stability.

In this book, students discover the nature of materials that make up themselves and their universe. The first step is to explore the general properties of matter and the connections among them. In Chapter 2, "Physical and Chemical Changes," students discover the four phases that matter can exist in, and they learn about the gas laws. Phase changes and their relationships to changes in heat energy are studied, and students learn to distinguish between a chemical property and a chemical change.

In the next chapter, students learn how to classify matter according to its makeup. As they learn about the various kinds of mixtures, they discover why solutions are different from other mixtures. Students also learn to use chemical symbols, formulas, and balanced equations to describe chemical reactions.

Chapter 4 begins by outlining the atomic model, from the early Greek concept to the modern atomic theory. Students explore the structure of the atom and discover the relationships among atomic number, isotope, mass number, and atomic mass.

Chapter 5 outlines the historical development of the Periodic Table up to the modern design, gives an in-depth look at the eight families of elements, and discusses how the periodic law explains the physical and chemical properties of elements.

The other *Prentice Hall Science* books integrated within MATTER: BUILDING BLOCK OF THE UNIVERSE are

Exploring the Universe
Ecology: Earth's Living Resources
Parade of Life: Animals
Dynamic Earth
Exploring Earth's Weather
Exploring Planet Earth
Ecology: Earth's Natural Resources
Cells: Building Blocks of Life
Evolution: Change Over Time
Parade of Life: Monerans, Protists, Fungi, and Plants

In addition to the Student Edition, Annotated Teacher's Edition, Teacher's Resource Package, and Laboratory Manual, the following media and technology components are integrated within the teaching strategies for MATTER: BUILDING BLOCK OF THE UNIVERSE:

MEDIA AND TECHNOLOGY

Interactive Videodisc
Science Vision: Chemical Pursuit

Videos/Videodiscs
Elements, Compounds, and Mixtures
Periodic Table and Periodicity
Chemical Bonding and Atomic Structure

Product-Testing Activities
Testing Paper Towels
Testing Yogurt

Testing Disposable Cups
Testing Bottled Water
Testing Glues

Transparencies
Molecules
Rutherford's Experiment
Structure of the Atom
Isotopes of Carbon
Energy Levels
The Periodic Table

Chemistry of Matter

Ch.1 Atoms and Bonding

Ch.2 Chemical Reactions

Ch.3 Families of Chemical Compounds

Ch.4 Chemical Technology

Ch.5 Radioactive Elements

Primary Themes Energy, Patterns of Change, Scale and Structure, Systems and Interactions, and Stability.

Students discover the answer to "What is bonding?" through an investigation of electrons, energy levels, and bonding. An in-depth look at ionic bonds, covalent bonds, and metallic bonds follows, along with opportunities to learn about and practice predicting bonds.

After being introduced to the law of conservation of mass in Chapter 2, students get a step-by-step lesson on balancing chemical equations—followed by plenty of opportunity to practice this skill. Also examined are the four types of chemical reactions, the classification of chemical reactions, and the rates of chemical reactions.

In order to better understand one of the most important families of chemical compounds—acids, bases, and salts—students first discover all about the process that produces them—making solutions. Another important family of chemical compounds—carbon compounds—are studied, including hydrocarbons and substituted hydrocarbons.

"Chemical Technology" focuses on what petroleum is and the process of separating it into useful parts. Students discover how polymers are formed, as well as explore the natural and synthetic polymers in their world.

The book ends with radioactive elements, where students learn the history of radioactivity and compare the three types of nuclear radiation. Nuclear reactions are discussed and nuclear fission and nuclear fusion are compared. Finally, the uses of radioactive substances—including advantages and disadvantages—are analyzed.

The other *Prentice Hall Science* titles integrated within CHEMISTRY OF MATTER are

Matter: Building Block of the Universe
Cells: Building Blocks of Life
Electricity and Magnetism
Dynamic Earth
Heat Energy
Motion, Forces, and Energy
Human Biology and Health
Ecology: Earth's Natural Resources
Chemistry of Matter
Exploring Planet Earth
The Nature of Science
Exploring the Universe
Parade of Life: Animals
Evolution: Change Over Time
Parade of Life: Monerans, Protists, Fungi, and Plants
Heredity: The Code of Life

MEDIA AND TECHNOLOGY

In addition to the Student Edition, Annotated Teacher's Edition, Teacher's Resource Package, and Laboratory Manual, the following media and technology components are integrated within the teaching strategies for CHEMISTRY OF MATTER:

Interactive Videodisc

Science Vision: Chemical Pursuits

Videos/Videodiscs

Chemical Bonding and Atomic Structure
Elements, Compounds, and Mixtures
Periodic Table and Periodicity
Acids, Bases, and Salts

Product-Testing Activities

Testing Antacids

Testing Cereals
Testing Nail Enamels

Transparencies

Periodic Table
Structure of the Atom
Molecules
Energy Levels
Ionic Bonding
Covalent Bonding
Acids, Bases, and pH Values

Electricity and Magnetism

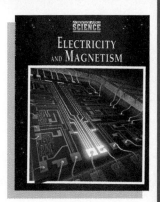

Ch.1 Electric Charges and Currents

Ch.2 Magnetism

Ch.3 Electromagnetism

Ch.4 Electronics and Computers

Primary Themes Energy, Scale and Structure, Systems and Interactions, and Stability.

Students begin their study of electricity by discovering what it is, how it is produced, and why it is so important. In doing so they come to understand the relationship between electric charges and atomic structure and how objects can acquire charge. After learning how charges can flow and discovering the relationships among voltage, resistance, and electric current, students explore electric circuits. They then apply this knowledge to explain how electric power is calculated and purchased.

Students discover what magnetism is, the properties that make a substance magnetic, and the significance of magnetism in everyday life. Along the way, they investigate the characteristics of magnetic fields and how magnetism is related to the atomic structure of a material. The Earth as a magnet is explored, as well as other sources of magnetism in the solar system, and students are introduced to how a compass works.

The relationship between electricity and magnetism is covered as students learn the principles behind creating magnetism from electricity and how electricity can be produced from magnetism. The understanding of these principles is then applied to devices such as electric motors, generators, and transformers. Lastly,

students become aware of the importance of electricity and magnetism in our lives by examining the devices that have brought the computer age and the electronic industry to where it is today.

The other *Prentice Hall Science* titles integrated within ELECTRICITY AND MAGNETISM are

Matter: Building Block of the Universe
Motion, Forces, and Energy
Chemistry of Matter
Exploring Earth's Weather
Heat Energy
Human Biology and Health
Parade of Life: Animals
Ecology: Earth's Natural Resources
Exploring the Universe
The Nature of Science
Exploring Planet Earth
Sound and Light
Dynamic Earth

In addition to the Student Edition, Annotated Teacher's Edition, Teacher's Resource Package, and Laboratory Manual, the following media and technology components are integrated within the teaching strategies for ELECTRICITY AND MAGNETISM:

MEDIA AND TECHNOLOGY

Interactive Videodiscs/CD ROM
Science Discovery: Static Electricity
Science Discovery: Circuits

Interactive Videodisc
Invention: Mastering Sound

Transparency
Series and Parallel Circuits

Heat Energy

Ch.1 What Is Heat?
Ch.2 Uses of Heat

After learning what heat is, students investigate three methods of heat transfer—conduction, convection, and radiation. In order to understand temperature and the difference between temperature and heat, a close look is taken at how energy and the motion of molecules are related. This concept is continued as students investigate how changes in temperature can provide a way to measure heat indirectly. To complete their understanding of heat, students explore heat and phase changes and thermal expansion.

The next chapter focuses on how heat is obtained, used, and controlled. Students explore various heating and cooling systems—including passive and active solar heating systems—and how insulation can prevent heat loss. Heat engines are investigated, with a look at external and internal combustion engines.

The book begins with a scenario predicting the impact of the greenhouse effect on life in the future and concludes with a closer look at some of the effects of thermal pollution on the environment.

The other *Prentice Hall Science* books integrated within HEAT ENERGY are

Ecology: Earth's Natural Resources
Exploring Earth's Weather
Exploring Planet Earth
Exploring the Universe
Human Biology and Health
Motion, Forces, and Energy
Matter: Building Block of the Universe
Electricity and Magnetism
Parade of Life: Animals

MEDIA AND TECHNOLOGY

In addition to the Student Edition, Annotated Teacher's Edition, Teacher's Resource Package, and Laboratory Manual, the following media and technology components are integrated within the teaching strategies for HEAT ENERGY:

Interactive Videodiscs
On Dry Land: The Desert Biome
The Solar Sea

Videodiscs
Atmospheric Series: Atmosphere and Winds
  and Air Currents
Atmospheric Series: Clouds and Precipitation
  and Global Winds

Product-Testing Activities
Testing Bottled Water
Testing Popcorn
Testing Disposable Cups

Transparencies
Three Methods of Heat Transfer
Solar Heating System

Sound and Light

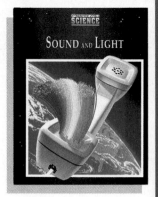

Ch.1 Characteristics of Waves   **Ch.3** Light and the Electromagnetic Spectrum

Ch.2 Sound and Its Uses   **Ch.4** Light and Its Uses

Primary Themes   Energy, Patterns of Change, Systems and Interactions, and Unity and Diversity.

An in-depth study of sound and light begins with a look at the nature of waves, how they are related to energy, and the difference between mechanical and electromagnetic waves. The characteristics of waves are explored as well as the different types of waves. Students discover how frequency and wavelength are related to wave speed, how to calculate the speed of a wave, and how various mediums can affect speed. They then go on to investigate basic wave interactions.

With a solid understanding of waves, students move on to study sound, beginning with a look at what it is and how it is produced. Properties of sound, such as pitch, intensity, and loudness are examined, as well as what happens when sound waves combine or are altered as in the Doppler effect. Students start to apply this knowledge as they study sound quality, the difference between music and noise, the many applications of sound—such as sonar and ultrasound—and how the body detects noise.

The focus shifts as students apply their knowledge of waves to the study of light. Electromagnetic waves are covered in depth, and regions of the electromagnetic spectrum are identified, along with the uses of electromagnetic waves at different frequencies. Students dis-cover how luminous objects produce light and investigate the question of whether light is a particle, a wave, or both. This background is used in the concluding chapter as students investigate the nature of a light ray, reflection and refraction of light, color, and the many applications of light.

The other *Prentice Hall Science* titles integrated within SOUND AND LIGHT are

> Exploring Planet Earth
> Motion, Forces, and Energy
> Matter: Building Block of the Universe
> Exploring the Universe
> Dynamic Earth
> Human Biology and Health
> Parade of Life: Animals
> Electricity and Magnetism
> The Nature of Science
> Exploring Earth's Weather
> Heat Energy
> Parade of Life: Monerans, Protists, Fungi,
>   and Plants

In addition to the Student Edition, Annotated Teacher's Edition, Teacher's Resource Package, and Laboratory Manual, the following media and technology components are integrated within the teaching strategies for SOUND AND LIGHT:

MEDIA AND TECHNOLOGY

Interactive Videodiscs/CD ROM
Science Discovery: Light and Lenses
Science Discovery: Waves
Science Discovery: Sound

Interactive Videodisc
Invention: Mastering Sound

Product-Testing Activity
Testing Nail Enamels

Transparencies
Wave Characteristics
Doppler Effect
Interference of Sound Waves
Human Anatomy: Ear
Electromagnetic Spectrum
Nearsightedness and Farsightedness
Human Anatomy: Eye

Motion, Forces, and Energy

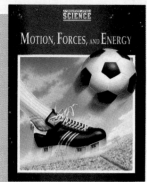

Primary Themes Energy, Patterns of Change, Scale and Structure, and Systems and Interactions.

Chapter 1 begins with a discussion of frames of reference. With a basic understanding of this concept, students move on to learn about measuring motion—speed and velocity, changes in velocity, and momentum.

After the question "What is force?" is answered, students move on to a study of friction and Newton's Laws of Motion. They then explore acceleration of falling objects due to gravity, air resistance, Newton's Law of Universal Gravitation, and the difference between weight and mass. Continuing in Chapter 3, the concept of fluid pressure is introduced, including a section on differences in pressure and how this has applications, from vacuum cleaners to our own respiratory system. The lesson concludes with a study of pressure and gravity, buoyancy, and fluids in motion.

Chapter 4 reveals the relationship between force, work, and distance, and the concept of power, leading to an exploration of simple and compound machines. In studying "Energy: Forms and Changes," students investigate the nature of energy, compare kinetic and potential energy, explore energy conversions and conservation of energy, and recognize the importance of energy in all physical processes.

The other *Prentice Hall Science* books integrated within MOTION, FORCES, AND ENERGY are

Exploring the Universe
Exploring Earth's Weather
Human Biology and Health
Matter: Building Block of the Universe
Parade of Life: Animals
Dynamic Earth
Electricity and Magnetism
Heat Energy
Chemistry of Matter
Sound and Light
Parade of Life: Monerans, Protists, Fungi, and Plants
Exploring Planet Earth

MEDIA AND TECHNOLOGY

In addition to the Student Edition, Annotated Teacher's Edition, Teacher's Resource Package, and Laboratory Manual, the following media and technology components are integrated within the teaching strategies for MOTION, FORCES, AND ENERGY.

Interactive Videodiscs/CD ROM
Science Discovery: Motion
Science Discovery: Forces
Science Discovery: Simple Machines

Interactive Videodiscs
Science Vision: ErgoMotion
Science Vision: AstroVision

Product-Testing Activities
Testing Disposable Cups
Testing Food Wraps

Transparencies
Hydraulic Lift
Forces and Airplane Flight

The Annotated Teacher's Edition provides you with simple, point-of-use guidance, whatever your teaching style or classroom needs.

The **Teacher's Guide** section, for example, gives you a basic overview of the integrated and thematic approaches to teaching science, including matrices that outline the themes found in each chapter, and the "big ideas" that exemplify each theme (see pages 62–63 of this *Teacher's Desk Reference* for an article on teaching thematically).

CHAPTER 1

Interactions Among Living Things

*ENERGY	• In general, food and energy in an ecosystem flow from the producers to the consumers, and finally to the decomposers. • Producers capture light energy, which cannot be used by consumers, and change it into food energy, which can be used by all living things. • Energy is lost from the chain at each feeding level in an ecosystem.
*EVOLUTION	• Species evolve in response to the challenges of their environment. • Interactions among organisms can affect the directions in which organisms evolve.
*PATTERNS OF CHANGE	• Each time a change occurs in an ecosystem, an adjustment in the ecosystem's balance is required. • Organisms change and are changed by their environment. • Changes in the environment may be slow or rapid and may involve individuals, species, or entire communities.
SCALE AND STRUCTURE	• The world can be divided into ecosystems, which in turn consist of smaller components. The world itself can be described as an ecosystem.
*SYSTEMS AND INTERACTIONS	• All of the living and nonliving things in an environment are interconnected. • The populations in a community interact in many different ways.
*UNITY AND DIVERSITY	• Although the Earth's ecosystems vary greatly, they all contain the same basic kinds of interactions.
STABILITY	• Ecosystems adjust in response to changes. • The interactions within an ecosystem are in a state of dynamic balance.

Chapter 1 — INTERACTIONS AMONG LIVING THINGS — CHAPTER PLANNING GUIDE*

SECTION	HANDS-ON ACTIVITIES	OTHER ACTIVITIES	MEDIA AND TECHNOLOGY
1–1 Living Things and Their Environment pages G12–G17 Multicultural Opportunity 1–1, p. G12 ESL Strategy 1–1, p. G13	**Student Edition** ACTIVITY (Doing): Identifying Interactions, p. G15 ACTIVITY (Discovering): Home Sweet Home, p. G17 **Laboratory Manual** Examining an Unknown Community, p. G7 **Teacher Edition** Observing Environments, p. G16d	**Activity Book** CHAPTER DISCOVERY: Exploring Relationships Among Living Things, p. G9 ACTIVITY: Populations in Your Community, p. G17 ACTIVITY: Carrying Capacity, p. G23 **Review and Reinforcement Guide** Section 1–1, p. G5	Interactive Videodisc/CD ROM Amazonia Interactive Videodisc Insects: Little Giants of the Earth Video Ecological Biology (Supplemental) Video/Videodisc Ecology: Communities (Supplemental) Ecology: Populations (Supplemental) English/Spanish Audiotapes Section 1–1
1–2 Food and Energy in the Environment pages G18–G23 Multicultural Opportunity 1–2, p. G18 ESL Strategy 1–2, p. G18	**Student Edition** ACTIVITY (Doing): Diet Delights, p. G19 ACTIVITY (Doing): Your Place in Line, p. G20 ACTIVITY BANK: Garbage in the Garden, p. G138 **Laboratory Manual** Investigating Relationships in an Ecosystem, p. G13 **Activity Book** ACTIVITY Observ...	**Activity Book** ACTIVITY: Find the Food Chain, p. G19 ACTIVITY: A Food Web, p. G25 **Review and Reinforcement Guide** Section 1–2, p. G7	Interactive Videodisc/CD ROM Virtual BioPark Video/Videodisc Ecology: Food Chains (Supplemental) Transparency Binder Food Web Energy Pyramid Courseware Food Chains and Webs (Supplemental) English/Spanish Audiotapes Section 1–2

Detailed guidance for implementing the program is offered for each chapter, both in the **Chapter Planning Guide** and **Chapter Preview** that immediately precede each chapter and throughout the chapter in the teaching notes surrounding the student pages.

The **Chapter Preview** provides an overview of the chapter that follows and, for each section, a thematic focus, performance objectives, and science terms. Additionally, opportunities for Discovery Learning are provided through teacher demonstrations and modeling.

The **Chapter Planning Guide** makes it easy for you to choose which print materials and media and technology components to integrate into your lesson. This guide also offers suggestions on when to integrate these materials into your lesson for maximum benefit.

Because the *Annotated Teacher's Edition* was designed to be teacher-friendly, the teaching notes surrounding the student pages virtually are self-explanatory. On the following four pages are some outstanding features that teachers have found particularly useful and exciting.

*E*ach chapter opener begins by pointing out specific opportunities for relating the chapter concepts to other sciences and disciplines. (See articles on integrating science in this *Teacher's Desk Reference*, pages 60–61 and 64–65.)

*T*his feature provides two kinds of discovery opportunities: one to discuss the impact of science and technology on society, and one for students to learn by interacting with their peers in groups. (See pages 70–75 in this *Teacher's Desk Reference* for an article on Cooperative Learning.)

CHAPTER 3
Exploring Earth's Biomes

INTEGRATING SCIENCE

This life science chapter provides you with numerous opportunities to integrate other areas of science as well as other disciplines into your curriculum. Blue-numbered annotations on the student page and integration notes on the teacher wraparound pages alert you to areas of possible integration.

In this chapter you can integrate earth science and meteorology (pp. 71, 85), earth science and geology (p. 71), life science and evolution (p. 72), earth science and maps (p. 74), life science and plants (p. 77), language arts (pp. 83, 91), earth science and oceanography (p. 88).

SCIENCE, TECHNOLOGY, AND SOCIETY/COOPERATIVE LEARNING

Economic, social, political, and environmental pressures threaten the existence of all six land biomes. Growing population exerts pressure to develop once untouched areas. Resources are used for economic growth. The result can be the destruction of biomes. Tropical rain forests provide a prime example of the pressures facing the Earth's biomes. People have cut into the rain forests to make more land available for farming and ranching. The main resource of the rain forest is harvested to meet the demand for wood. Teak and other forest trees are made into furniture and other goods for export.

INTRODUCING CHAPTER 3

DISCOVERY LEARNING

▶ *Activity Book*
You may want to begin your teaching of the chapter by using the Chapter 3 Discovery Activity from your *Activity Book*. Using this activity, students will discover that organisms and climate are characteristic of different environments.

USING THE TEXTBOOK

Have students observe the photo on page G68.
• **What do you predict is happening in the picture?** (Students might suggest the lions are scanning the grasslands in search of prey.)
• **In what kind of environment do the lions live?** (Students might identify the environment as grasslands or plains.)
• **What other animals would you expect to see in this environment?** (Accept all log-

ical answers. Students may name a such as antelopes that the lion pre

Point out to students that lion the open plains in Africa. Have dents read the chapter introductio students identify the different a that live on the African plains. 1 dents that these animals interact w another and rely on the environm their food sources.
• **What do zebras and gazelles eat?** grasses.)

68 ■ G

*E*ach chapter introduction refers you to a Discovery Learning activity in the *Activity Book*. In addition, each *Prentice Hall Science* book begins with a Discovery Activity. There is a "Problem Solving" feature in virtually every chapter, and there is at least one "Activity (Doing)," "Activity (Calculating)," "Activity (Reading)," "Activity (Writing)," or "Activity (Thinking)" feature in each section of every chapter. Teaching strategies for each of these activities are integrated in the *Annotated Teacher's Edition* on the same pages as the student pages to which they refer. (See pages 66–67of this *Teacher's Desk Reference* for an article on Discovery Learning.)

xploring Earth's iomes

de for Reading

you read the following ons, you will be able to

Biogeography
■ Explain how plants and animals move from one area to another.

Tundra Biomes
■ Describe the characteristics of tundra biomes.

Forest Biomes
■ Compare the characteristics of three forest biomes.

Grassland Biomes
■ Describe the characteristics of grassland biomes.

Desert Biomes
■ Describe the characteristics of desert biomes.

Water Biomes
■ Identify and describe the characteristics of the major water biomes.

Night falls quickly on the vast Serengeti Plain of East Africa, as if a black velvet curtain has suddenly been drawn over the land. The scattered acacia trees and the great herds of zebras, wildebeests, and gazelles that graze on the Plain during the day disappear in the sudden darkness.

Safe in camp, you sit in your tent and listen to the mysterious sounds of the night. Nearby, a family of zebras snorts and stomps, startled perhaps by the rumble of distant thunder. The wildebeests, or gnus, stir and shuffle as they settle down for the night. And then the lions begin to roar. The wild music of the lions sends chills down your spine.

Lions, zebras, wildebeests, and gazelles do not live everywhere in Africa. They inhabit only the open plains, or savannas, with few or no trees and plenty of grass. Different animals live in the steamy jungles, which have many trees but not much grass. As you will discover in the pages that follow, animal and plant populations are not the same from place to place. They vary because different areas of the Earth have different climates. Climate conditions play a large role in determining where organisms make their homes.

Journal *Activity*

You and Your World Perhaps you have camped in a state or national park, spent the night in a tent in your own backyard, or imagined what it would be like to camp out. In your journal, describe your experiences, whether they are actual or imagined.

and watchful, a group of lionesses scan the grassy African plain for their prey.

G ■ 69

Some countries have designed large-scale relocation programs to reduce overcrowding or to equalize the distribution of land. These programs have had devastating effects on the rain forest. They have also affected the cultures of the original rain-forest inhabitants. These people's traditional way of life may be lost, and they are exposed to new and often deadly diseases brought in by newcomers. Building projects designed to promote the development of Third World countries also have sped up the destruction of the rain forests.

The entire world has a stake in what happens to the rain forests. Deforestation has disrupted the carbon dioxide–oxygen cycle, which in turn affects the global climate. Species of plants and animals whose value is still unknown are destroyed. Therefore, the conservation and preservation of the rain forest and other biomes are a worldwide responsibility.

Cooperative learning: Using preassigned lab groups or randomly selected

• Prepare a three-minute segment for a television news magazine focusing on the plight of the rain forest. Groups should write scripts and prepare visuals for their segments.
• Illustrate a biome before and after the removal of a natural resource, for example, oil or timber, from the biome.

See Cooperative Learning in the *Teacher's Desk Reference.*

JOURNAL ACTIVITY

You may want to use the Journal Activity as the basis of a class discussion. As students discuss their camping experiences, help them identify ways that they can protect the environment, such as by cleaning up their litter and staying on designated trails. Students should be instructed to keep their Journal Activity in their portfolio.

*S*tudent Text

hy do you think they live on the African ains? (They feed on its grasses.)
What do lions eat? (Other animals, eat.)
Why do you think the lion lives on the rican plains? (It feeds on the grazing animals there.)
Point out that the grasslands are only e type of environment in which animals and plants live.
Where do you think deer live? (Students ight suggest the woods or the forest.)

• **Where do polar bears live?** (Accept such responses as in the Arctic or in Alaska.)
• **Where would you expect to see a cactus?** (The desert.)
Help students conclude that different environments support different types of animals and plants.

G ■ 69

*T*he Journal Activity signals an opportunity for helping students develop their portfolios as a means of Alternative Assessment. "Alternative Assessment" and "Keeping a Portfolio" teaching strategies are also included in each Chapter Review. (See pages 76–77 in this *Teacher's Desk Reference* for an article on Authentic Assessment.)

Every section begins with a suggestion for bringing out the richness and diversity of different cultures that students may bring with them to the classroom. This also provides opportunities for highlighting the contributions made to science and humanity in the past and present by different cultures. (See pages 84-85 of this *Teacher's Desk Reference* for an article on Multiculturalism.)

Every section contains a strategy for teaching ESL (English as a Second Language) students. (See pages 86-87 of this *Teacher's Desk Reference* for an article on English as a Second Language.)

3-1 Biogeography

MULTICULTURAL OPPORTUNITY 3-1

Many anthropologists believe that cultures developed by way of migration. Approximately 30,000 years ago, when Siberia and Alaska were connected by a land bridge, humans migrated from the regions of Asia and spread south across the Canadian plains. Encourage interested students to explore further these theories of cultural dispersion.

ESL STRATEGY 3-1

Tell students that in English, as perhaps in their native languages, many words are derived from ancient Greek. Go over the meanings of the following word roots and affixes.

bio: life
geo: earth
eco: habitat, environment
graphy: description
logy: study

Ask students what they think the section heading Biogeography means. Ask them to deduce meanings for biology, biography, geography, geology, ecology. Are there other words they know that use the Greek roots and affixes?

Guide for Reading

Focus on these questions as you read.
▶ How do plants and animals disperse?
▶ What are the six major land biomes?

3-1 Biogeography

You are an explorer. In this chapter, you a[re go]ing on a trip around the world. Your trip will [take] you from the cold, barren lands surrounding [the] North Pole to the dense jungles near the equ[ator,] and even into the depths of the oceans. As yo[u trav]el, you will discover many strange and wonde[rful] plants and animals. You will find that the kin[ds of] plants and animals change as you move from [place] to place on your journey around the world.

The study of where plants and animals liv[e] throughout the world (their distribution) is c[alled] **biogeography.** Biogeographers, then, are inter[ested] in ecology, or the study of the relationships a[mong] plants, animals, and their environment.

The kinds of animals that live in an area [depend] largely on the kinds of plants that grow there[. Do] you know why? Animals rely on plants as one [source] of food. For example, zebras eat mostly grass. [They] would have a difficult time finding enough fo[od in a] jungle, where grass is scarce. But grassy plains [are a] good habitat, or living place, for zebras. Plain[s are] also a good habitat for lions. Why? Lions are [meat] eaters (carnivores) that hunt plant eaters (herbivores), such as zebras, for food.

Figure 3-1 *The gray-headed albatross makes its nest o[f] mud and grass on small wind-swept islands in the Sout[hern] Hemisphere (bottom left). Meerkats live in Africa's Kala[hari] desert (top). In what kind of African habitat would you e[xpect to] find a lowland gorilla (bottom right)?* ❶

70 ■ G

TEACHING STRATEGY 3-1

FOCUS/MOTIVATION

Prepare ahead of time by collecting several pictures of plants associated with different biomes. For example, you might collect and display pictures of palm trees along a beach, grasses on a prairie, trees in a co[niferous] [forest], and cacti in a desert. When [displ]aying the pictures, challenge studen[ts to i]dentify characteristics of the environm[ent] in which the plants would grow.

• **Are the plants shown likely to grow in the same area?** (Answers may vary. Lead students to suggest that the plants naturally grow in different areas.)

◈ Media and Technology

Students can explore a complete desert ecosystem, investigating those aspects of the desert that interest them, by beginning the In[tera]ctive Videodisc called The Dese[rt.]

CONTENT DEVELOPMENT

As you introduce the concept o[f bio]geography, help students identify [the in]terrelationship of biology, the st[udy of] life and of geography, and the st[udy of] Earth's surface and features. Intr[oduce] the term *biogeographers* and have st[udents] predict the kinds of topics these [scien]tists would study. You might offer t[he fol]lowing suggestions: animals that liv[e in an] area, plants that live in an area, the c[limate,] and other characteristics of an are[a.]

70 ■ G

The teaching strategies are based on an easy-to-follow lesson plan format that includes Focus/Motivation, Content Development, Guided Practice, Independent Practice, Enrichment, Reinforcement/Reteaching, and Closure.

Suggestions for using the print and media and technology components in the Integrated Learning System are built into the lesson plan format at the point of use.

turn, the plant life in an area is determined
by climate. Climate describes the average
tions of temperature and precipitation (rain,
sleet, hail) in an area over a long period of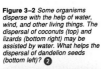
Trees grow tall and dense in warm, rainy cli-
, especially if the days are long and there is
of sunlight throughout the year. Fewer trees
in cold, dry climates, where the short days of
arrive early and stay late.

...ersal of Plants and Animals

addition to studying where plants and animals
biogeographers also study why plants and ani-
spread into different areas of the world. The
ment of living things from one place to anoth-
called **dispersal**. Plants and animals disperse in
ways. For example, about 50 million years ago,
s evolved in North America. During prehistoric
, the sea level dropped and a land bridge
ed between Alaska and Siberia, which is in Asia.
es soon moved westward across this natural land
e into Asia. Over many thousands of years,
s dispersed all across northern Asia and into
pe.

ometimes plants and animals disperse with
—from water, wind, and even people. Certain
ls, for example, have spread from island to is-
on floating branches. Some seeds, such as co-
ts, also reach new places by floating on water.
n microorganisms, the spores of fungi, dande-
seeds, baby spiders, and many other small, light
nisms may be carried by the wind to new places.

Figure 3–2 *Some organisms
disperse with the help of water,
wind, and other living things. The
dispersal of coconuts (top) and
lizards (bottom right) may be
assisted by water. What helps the
dispersal of dandelion seeds
(bottom left)?*

G ■ 71

BACKGROUND INFORMATION
DISPERSAL

Dispersal may be active or passive. Active dispersal occurs when an animal
moves by its own energy from one place
to another—a bird flying or a fish swim-
ming, for example. Passive dispersal occurs when an organism is carried from
place to place by wind, moving water, animals, or other physical factors. An example of passive dispersal is an acorn
buried by a squirrel.

*H*igh-interest tie-ins encourage
integration of background information, science and other disciplines, notes on ecology, facts
and figures, etc., into your lesson
plan at the point of use.

' use of the environment, dispersal of
ts and animals. Have students read
section to confirm their predictions.

● ● ● **Integration** ● ● ● ●

se the text copy on climate to inte-
e meteorology into your lesson.

Use the information about the land
ge between Alaska and Sib..ria to in-
ate geology into your les...

GUIDED PRACTICE

Skills Development

Skill: Classifying

Take a five-minute field trip in or near
the school grounds to see what kinds of
seeds you can discover. Classify the seeds
according to methods of dispersal based
on their shape and other adaptations.
• **What are some characteristics of seeds
that "hitchhike"?** (Spines that attach to fur
or clothing.)

• **Of seeds that disperse by means of water?** (Have hard coats, float easily.)
• **Of seeds that disperse by means of the
wind?** (Very light, generally have large
surface areas to capture the wind.)

ENRICHMENT

Encourage interested students to co-
operatively prepare a class report on the
evidence that supports the hypothesis that
a land bridge connected Alaska and
Siberia. Use the report to stimulate class
discussion on the hypothesis.

G ■ 71

*O*pportunities for integrating concepts from other sciences and
disciplines are built into the lesson plan at the point of use. These
opportunities are also indicated on the student pages by a numbered
blue dot that refers you to the Annotation Key for source information. There is another opportunity for teaching an integrated approach in every chapter with the "Connections" feature.

	A — The Nature of Science	B — Parade of Life: Monerans, Protists, Fungi, and Plants
Energy	All physical phenomena and interactions involve energy.	All organisms, autotrophs and heterotrophs alike, need energy to live.
Evolution	• Our understanding of the world around us has changed over time. • Scientific knowledge and theories are constantly being modified.	Evolutionary relationships are the basis for modern classification of living things.
Patterns of Change	Data collected through experimentation and observation can lead to the development of new theories about the way the universe works.	• Plants, fungi, and protists may change form as they pass through the stages of their life cycle. • Organisms change in predictable ways in response to their environment.
Scale and Structure	• Scientists seek answers to questions about the world on both microscopic and macroscopic levels. • Science is traditionally divided into life, earth, and physical branches.	The functioning of individual structures contributes to the functioning of the organism as a whole.
Systems and Interactions	Scientists all over the world can communicate with one another because they investigate the natural world according to a basic method known as the scientific method and use a standard system of measurement known as the metric system.	Living things interact with one another and with their environment.
Unity and Diversity	The different branches of science have the same basic goal: to increase human understanding of the world.	Although living things vary greatly in size, complexity, and structure, they share certain characteristics and interact with their surroundings in the same basic ways.
Stability	• Scientists use the same basic approach to study phenomena—the scientific method. • Scientific theories and laws usually remain fairly stable because they are based on consistent evidence.	Living things are involved with cycles of matter that balance the different forms of substances in the environment.

C *Parade of Life: Animals*	D *Cells: Building Blocks of Life*	**Theme Trace for Prentice Hall Science**
Animals obtain energy by eating other living things.	Special cell processes capture light energy, change light energy to a form that living things can use, and break down food to release the energy needed to power cell functions.	**Energy**
Vertebrates evolved from invertebrate ancestors. Mammals and birds evolved from reptiles, which evolved from amphibians, which evolved from fishes.	Cells have evolved into many specialized forms.	**Evolution**
In animal evolution there are trends toward increasing complexity and specialization.	Cell growth and development relate to the life cycles of living things.	**Patterns of Change**
The way animals perform their life functions can be studied at different levels of structure, from cells to organs to organ systems to organisms to ecosystems.	• Organisms are made of cells, which in turn consist of many parts. • From largest to smallest, the basic levels of organization of a complex organism are cell, tissue, organ, and organ system.	**Scale and Structure**
• Animals affect and are affected by the living and nonliving parts of their environment. • An animal's cells and organs interact to maintain the life of the animal.	• The parts of a cell work together to carry out the cell's life functions. • The complementary interaction of photosynthesis and respiration forms the carbon and oxygen cycles.	**Systems and Interactions**
Although animals share the same basic needs, the structures and methods for carrying out life functions vary enormously.	Despite their many differences in size, shape, and function, all cells with a nucleus have the same basic structure.	**Unity and Diversity**
All animals perform basic life functions that allow them to maintain a stable internal environment.	• Certain cell processes help to ensure that the genetic material remains the same in parent cells and daughter cells. • Energy can be neither created nor destroyed. It can, however, be converted from one form to another.	**Stability**

	E Heredity: The Code of Life	**F** Evolution: Change Over Time
Energy		•Earth's living things, past and present, can be classified according to how they obtain energy. •Ultraviolet radiation may have provided the energy required to form the chemicals of life on the early Earth.
Evolution	Changes in genes, or mutations, may be passed from one generation to the next.	•Fossils provide clear evidence that living things have changed over time. •All living things have evolved from other living things. •The Earth's climate, landforms, and living things have changed over time.
Patterns of Change	•Traits are passed from parent to offspring. •Changes in the genetic code may produce changes in proteins.	•As a result of natural selection, some traits become more common while others disappear. •The first living things were quite simple; over time, increasingly complex organisms evolved. •At several points in the Earth's history, large numbers of species have become extinct at the same instant in geological time.
Scale and Structure	Genes are located on chromosomes, which are made up of coiled strands of DNA.	Evidence of evolution is present on many levels, ranging from whole fossil organisms to molecules within living organisms to radioactive atoms inside fossils.
Systems and Interactions	•Genes may affect the expression of other genes. •An organism's appearance and behavior reflects the ways its genes interact.	Interactions among living things, such as predation and competition, may lead to evolutionary changes or extinctions.
Unity and Diversity	Variations in the genetic code account for the diversity of traits among organisms.	•Although the members of a species are very similar, there are some variations among individuals. •Many different kinds of living things have existed on the Earth during its long history.
Stability	Genes are passed from one generation to the next.	•The basic processes involved in evolution remain the same throughout time. •The interactions of autotrophs and heterotrophs maintain the current composition of the Earth's atmosphere.

G Ecology: Earth's Living Resources	H Human Biology and Health	Theme Trace for Prentice Hall Science
Energy in an ecosystem flows from producers to consumers to decomposers.	Like all living things, humans need energy to perform their life activities.	Energy
Species evolve in response to living and nonliving factors in their environment.	The complexity of human organ systems reflects the challenges that the human species has met during the course of evolution.	Evolution
Matter changes form as it cycles through the environment.	The pattern of function of organ systems involves responses to internal and external changes.	Patterns of Change
The world can be divided into ecosystems, which in turn can be divided into smaller and smaller components.	The human body shows cell, tissue, organ, and organ system levels of organization.	Scale and Structure
All of the living and nonliving things in an environment are interconnected.	Organ systems work together in harmony to enable the body to adapt to changes in its internal and external environment.	Systems and Interactions
The Earth's ecosystems vary greatly, but all are affected by the same patterns of change and cycles of matter.	Organ systems differ in structure but are similar in that they are composed of tissue and work to perform a specific life function.	Unity and Diversity
The interactions within an ecosystem are in a state of dynamic balance.	The organ systems work together to maintain a stable internal environment.	Stability

Theme Trace for Prentice Hall Science	**I** *Exploring Planet Earth*	**J** *Dynamic Earth*
Energy	The energy of the sun makes life on Earth possible, influences the atmosphere and oceans, and drives the water cycle.	Energy resulting from heat and pressure within the Earth causes faulting and folding, produces earthquakes and volcanoes, moves continents, and changes rocks.
Evolution	The Earth's atmosphere, oceans, and crust have changed greatly over time.	The Earth's surface has changed over time and will continue to change as the result of dynamic processes.
Patterns of Change	The atmosphere, hydrosphere, and lithosphere undergo changes that are determined by living and nonliving factors.	The Earth's crust is changed in specific ways by a number of subsurface and surface activities.
Scale and Structure	•The Earth's surface is made up of the atmosphere, hydrosphere, and lithosphere, which in turn are divided into smaller components. •The Earth consists of the core, mantle, and crust.	The surface of the Earth consists of lithospheric plates, which are composed of rocks, which are made up of minerals. Some of the activities that change the Earth's crust are obvious; others are subtle.
Systems and Interactions	The processes that cycle water and many atmospheric gases through the environment involve the interactions of the atmosphere, hydrosphere, and lithosphere.	•Plate movements result in earthquakes, volcanoes, and the formation of mountains and valleys. •Rocks are changed into other kinds of rocks and into soil by processes such as weathering and metamorphism.
Unity and Diversity	•The layers of the Earth have different properties. •The different ocean life zones have different kinds of conditions and organisms. •The layers of the atmosphere, which are distinguished by different compositions and temperatures, form a protective blanket around the Earth.	•Forces within the Earth are manifested in many different ways—earthquakes, folding, and plate movement, to name a few. •Although rocks and minerals vary greatly in their physical properties, they are all made up of basically the same elements.
Stability	Various processes exist that maintain the balance of gases in the atmosphere, keep constant the amount of salts in the ocean, and stabilize the Earth's rotation and internal temperature and pressure.	•The downward force of the crust is balanced by the upward force of the mantle. •The processes that destroy crust are balanced by the processes that make crust.

K Exploring Earth's Weather	L Ecology: Earth's Natural Resources	Theme Trace for Prentice Hall Science
The Earth receives radiant energy from the sun. The unequal distribution of this energy produces weather.	•Earth's natural resources provide sources of energy. •Conservation of nonrenewable energy resources is everybody's responsibility.	**Energy**
•Living things have evolved adaptations for survival in a variety of climatic conditions. •The Earth's climate has undergone many changes during the planet's history.		**Evolution**
•Changes in atmospheric conditions (heat energy, air pressure, winds, and moisture) cause differences in weather. •Changes in climate are due to global factors.	•Energy can be changed from one form to another. •Human activities cause pollution, depletion of resources, and other changes in the environment.	**Patterns of Change**
The Earth's oceans and landmasses can be divided into climatic zones and smaller climatic regions.	Resources can be used (and misused) on personal, community, and global levels.	**Scale and Structure**
Many interacting factors—including latitude, elevation, continental drift, and the sun's energy output—shape weather and climate.	The ways in which humans interact with natural resources have powerful positive and negative effects on the environment.	**Systems and Interactions**
•Weather in a specific area changes from day to day, but climate (weather over a long period of time) remains fairly constant. •There are many different climate zones on Earth, each with its own characteristic temperature, precipitation, and set of living things.	•Natural resources, renewable as well as nonrenewable, are important to humans and can be used wisely or foolishly. •The three major types of pollution—land, air, and water—all damage the environment.	**Unity and Diversity**
Earth's climate remains relatively unchanged over thousands of years.	•Energy from the sun is constant and unlimited. •The Earth's supply of renewable resources is constantly replenished by natural processes. •The wise use of natural resources will ensure a supply for the future. The unwise use will alter Earth's natural balance.	**Stability**

	M **Exploring the Universe**	**N** **Matter: Building Block of the Universe**
Energy	•Energy is released from a star as a result of nuclear fusion. •Atmospheric gases trap energy. •Different forms of energy account for the formation of the universe and for the various phenomena that occur in it.	•All matter consists of atoms, which are in constant motion due to their energy. •The energy content of an electron determines its position in the electron cloud. •Any chemical change in matter is accompanied by a change in energy.
Evolution	Stars and solar systems slowly evolve, or change over time.	
Patterns of Change	•Stars undergo a series of changes from birth to death. •The movements of the Earth, moon, and sun cause phenomena such as eclipses, tides, day and night, and seasons.	•Matter changes phase with either a loss or gain of heat energy. •Atoms of elements chemically combine in definite ratios to form compounds. •The physical and chemical properties of the elements are periodic functions of their atomic number.
Scale and Structure	The universe is composed of galaxies, which in turn are composed of stars. Our star system is composed of the sun, planets, moons, and other celestial bodies.	The basic building block of matter is the atom. The arrangement of electrons in the atom determines its properties.
Systems and Interactions	•Eclipses are produced by the interaction of the sun, moon, and Earth. •All objects in the universe exert a gravitational pull on one another. The results of these attractions include tides, rotation, revolution, and eclipses.	Physical changes are characterized by interactions of matter that do not produce new substances. During chemical changes, atoms are rearranged, thus producing new and different substances.
Unity and Diversity	Stars are capable of producing their own light and heat energy. Planets and moons are not. However, these objects all exert a gravitational pull on one another that maintains the structures of star systems.	•Matter is made up of atoms, which in turn are made up of electrons, neutrons, and protons. •The number of protons in an atom determines its identity and its physical and chemical properties. •Matter can exist as elements, compounds, mixtures, and solutions.
Stability	•The magnitudes of the gravitational force between objects is in proportion to their size and distance. •The planets in the solar system have changed little over the past few billion years.	•The mass of an object is constant unless matter is added or removed. •The structure of an atom of an element determines its physical and chemical properties, which can be predicted.

O *Chemistry of Matter*	P *Electricity and Magnetism*	**Theme Trace for Prentice Hall Science**
•Energy binds the nucleus of an atom together. •Energy is always involved in chemical bonding. •A chemical reaction can either give off energy or absorb energy.	•Energy is required to make electrical charges move and flow and to overcome magnetic forces. •Energy can be converted from one form to another.	*Energy*
Petroleum is believed to have formed from the remains of organisms that were buried and subjected to tremendous heat and pressure over millions of years.	•The Earth's magnetic field has changed over time. •The field of electronics has undergone rapid evolution.	*Evolution*
An atom can undergo a change in its nucleus (radioactivity) or in its electron configuration (bonding). In either case, a new substance is formed.	Like charges and poles repel; unlike charges and poles attract.	*Patterns of Change*
The arrangement of the electrons in an atom is unique to that atom and determines its chemical behavior.	The electric and magnetic properties of atoms give rise to electricity and magnetism, and allow the application of these two phenomena in the field of electronics.	*Scale and Structure*
Atoms will react with one another in a variety of ways, including the formation of solutions and compounds and the emission of radioactivity. These interactions are often dependent on temperature, pressure, concentration, surface area, and the presence of catalysts.	The motion of a charged particle is altered by magnetic forces in a magnetic field. A magnetic field will exert a force on a wire carrying current.	*Systems and Interactions*
•All chemical reactions, regardless of type, involve the production of new substances and a change in energy content. •All organic compounds contain carbon. •All nuclear reactions involve a change in the number of protons in the nucleus.	•Electric current can move in only one direction or can change direction repeatedly. Electric current can follow a single path or several different paths. •Electromagnets can be used to convert electrical energy to mechanical energy and mechanical energy to electrical energy.	*Unity and Diversity*
•Mass is always conserved in chemical reactions. •The result of radioactive decay is a stable nucleus.	•Charge is neither created nor destroyed; it is merely transferred. •Magnetic and electric forces behave in predictable ways.	*Stability*

	Q *Heat Energy*	**R** *Sound and Light*
Energy	Heat energy is caused by the internal motion of molecules.	Waves, such as those of light and sound, carry energy from one place to another through either a medium or a vacuum.
Evolution		• Organisms have evolved in ways that make them sensitive to the sun's electromagnetic waves. • Increased scientific knowledge leads to the evolution of new technologies.
Patterns of Change	• A phase change is always accompanied by a change in heat energy. • Heat moves from a warmer object to a cooler object. • Heat energy can be converted into other forms of energy.	• The speed of a wave is determined by its frequency and wavelength. These three properties also determine various characteristics of waves. • Sound and light waves taken in through sense organs are converted into electrical impulses interpreted by the brain.
Scale and Structure	Matter is made up of molecules, whose constant motion underlies all heat-related phenomena.	Any wave is a disturbance that carries energy by altering electric and magnetic fields or by changing the motion of particles of matter.
Systems and Interactions	The flow of heat energy between objects or areas accounts for the heating of the Earth, weather, and heating and cooling systems.	• The speed of a wave depends on the medium through which it travels. • Waves interact by reflection, refraction, diffraction, and interference.
Unity and Diversity	The various forms of heat transfer involve the flow of heat energy from an area of more heat energy to an area of less heat energy.	• Waves carry energy from one place to another. Sound waves are mechanical waves; light waves are electromagnetic waves. • The various regions of the electromagnetic spectrum arise from particular sources and have distinct applications.
Stability	During a phase change, the average kinetic energy of molecules does not change. During a temperature change, the average kinetic energy of molecules changes in a predictable way.	The speed of a wave in a given medium (or a vacuum) is constant.

S
Motion, Forces, and Energy

Theme Trace for Prentice Hall Science

•Energy is the ability to do work, or exert a force through a distance. •Energy can exist in a number of forms and can be converted from one form to another.	***Energy***
All forms of life and the changes they undergo require energy.	***Evolution***
•The application of energy starts, stops, or changes the direction/speed of motion of an object. •Bernoulli's principle and Archimedes' principle describe the nature and effect of forces in fluids. •Machines change the size of the effort force and the distance over which it is exerted. They can also change the direction of the force. •Energy can be changed from one form to another.	***Patterns of Change***
The movement of all objects, from subatomic particles to galaxies, can be described in terms of the application of forces and the formulas of motion.	***Scale and Structure***
•All objects exert a force. Balanced and unbalanced forces account for the behavior of moving objects. •Machines are used to change the size, direction, and effect of a force. •Energy can be changed from one form to another.	***Systems and Interactions***
•Newton's three laws of motion describe all aspects of an object's motion (above the subatomic level). •All objects exert a force, although the nature of that force may vary. •Energy, regardless of its form, is the ability to do work.	***Unity and Diversity***
•Momentum and energy are conserved in any system. •All objects in the universe attract one another by the force of gravity. •The description of an object's motion depends on the frame of reference of the observer. •Work output can never be greater than work input.	***Stability***

PRENTICE HALL SCIENCE
Sample Courses of Study

The Prentice Hall Science program can be used to arrange topics into many courses of study for grades six through nine. Due to the tremendous flexibility of Prentice Hall Science, it is also possible to teach fewer topics in greater depth than ever before. (See the professional article "Less Is More.") The following charts provide several examples of how Prentice Hall Science can be configured to meet a wide variety of curriculum needs. Keep in mind that these are only suggestions and that you can customize Prentice Hall Science to your individual classroom needs. In addition, you always have the option to reconfigure Prentice Hall Science at any time in any way you deem appropriate.

The suggested courses of study listed here are only a few of the many configurations possible with Prentice Hall Science.

Chart 1 This is an example of the books you might choose for a **general integrated science** curriculum for grades six through nine.

GENERAL INTEGRATED SCIENCE	GRADE 6	GRADE 7	GRADE 8	GRADE 9
A The Nature of Science	■	■	■	■
B Parade of Life: Monerans, Protists, Fungi, and Plants (L)		■		
C Parade of Life: Animals (L)	■			
D Cells: Building Blocks of Life (L)				■
E Heredity: The Code of Life (L)				■
F Evolution: Change Over Time (L, E)				■
G Ecology: Earth's Living Resources (L)	■			
H Human Biology and Health (L)			■	
I Exploring Planet Earth (E)		■		
J Dynamic Earth (E)				■
K Exploring Earth's Weather (E)			■	
L Ecology: Earth's Natural Resources (E)		■		
M Exploring the Universe (E)	■			
N Matter: Building Block of the Universe (P)	■			
O Chemistry of Matter (P)		■		
P Electricity and Magnetism (P)			■	
Q Heat Energy (P)			■	
R Sound and Light (P)	■			
S Motion, Forces, and Energy (P)				■

(L) – Life Science, (E) – Earth Science, (P) – Physical Science

Chart 2 This is an example of the books you might choose for a **spiral** approach to a **general science** curriculum for grades six through nine.

GENERAL INTEGRATED SCIENCE—SPIRAL APPROACH	GRADE 6	GRADE 7	GRADE 8	GRADE 9
A The Nature of Science	■	■	■	■
B Parade of Life: Monerans, Protists, Fungi, and Plants (L)		■		
C Parade of Life: Animals (L)	■			
D Cells: Building Blocks of Life (L)				■
E Heredity: The Code of Life (L)				■
F Evolution: Change Over Time (L, E)				
G Ecology: Earth's Living Resources (L)		■		
H Human Biology and Health (L)			■	
I Exploring Planet Earth (E)	■			
J Dynamic Earth (E)		■		■
K Exploring Earth's Weather (E)			■	
L Ecology: Earth's Natural Resources (E)		■		
M Exploring the Universe (E)			■	
N Matter: Building Block of the Universe (P)	■			■
O Chemistry of Matter (P)				■
P Electricity and Magnetism (P)		■		
Q Heat Energy (P)	■			■
R Sound and Light (P)			■	
S Motion, Forces, and Energy (P)	■			■

(L) – Life Science, (E) – Earth Science, (P) – Physical Science

Chart 3 This is an example of the books you might choose for a sixth grade **general science** curriculum and for a **life, earth,** and **physical science** approach for grades seven through nine.

GENERAL INTEGRATED/ LIFE, EARTH, PHYSICAL	GRADE 6	GRADE 7	GRADE 8	GRADE 9
A The Nature of Science	■	■	■	■
B Parade of Life: Monerans, Protists, Fungi, and Plants (L)		■		
C Parade of Life: Animals (L)	■			
D Cells: Building Blocks of Life (L)		■		
E Heredity: The Code of Life (L)		■		
F Evolution: Change Over Time (L, E)		■	■	
G Ecology: Earth's Living Resources (L)	■			
H Human Biology and Health (L)		■		
I Exploring Planet Earth (E)			■	
J Dynamic Earth (E)			■	
K Exploring Earth's Weather (E)			■	
L Ecology: Earth's Natural Resources (E)			■	
M Exploring the Universe (E)	■			
N Matter: Building Block of the Universe (P)	■			
O Chemistry of Matter (P)				■
P Electricity and Magnetism (P)				■
Q Heat Energy (P)				■
R Sound and Light (P)	■			
S Motion, Forces, and Energy (P)				■

(L) – Life Science, (E) – Earth Science, (P) – Physical Science

Chart 4 Use the following chart along with the information in the *Prentice Hall Science Annotated Teacher's Edition* to integrate themes into your science curriculum. Integrating and teaching thematically can be accomplished no matter what configuration of *Prentice Hall Science* is used. However, if thematic teaching is your primary goal, this chart identifies the themes that most commonly occur in *Prentice Hall Science* and links them to the most appropriate topics.

GENERAL INTEGRATED SCIENCE—THEMATIC APPROACH	ENERGY	EVOLUTION	PATTERNS OF CHANGE	SCALE AND STRUCTURE	SYSTEMS AND INTERACTIONS	UNITY AND DIVERSITY	STABILITY
A The Nature of Science				■	■	■	
B Parade of Life: Monerans, Protists, Fungi, and Plants (L)		■	■	■		■	
C Parade of Life: Animals (L)		■	■	■		■	
D Cells: Building Blocks of Life (L)	■	■		■	■		■
E Heredity: The Code of Life (L)		■	■		■	■	
F Evolution: Change Over Time (L, E)		■	■		■	■	
G Ecology: Earth's Living Resources (L)	■	■	■		■	■	
H Human Biology and Health (L)	■			■	■		■
I Exploring Planet Earth (E)	■		■	■	■		
J Dynamic Earth (E)	■	■	■	■			■
K Exploring Earth's Weather (E)	■		■		■		
L Ecology: Earth's Natural Resources (E)	■			■	■		■
M Exploring the Universe (E)	■	■		■	■	■	
N Matter: Building Block of the Universe (P)			■	■	■		■
O Chemistry of Matter (P)	■		■	■	■		■
P Electricity and Magnetism (P)	■			■	■		■
Q Heat Energy (P)	■						■
R Sound and Light (P)	■		■		■	■	
S Motion, Forces, and Energy (P)	■		■	■	■		

(L) – Life Science, (E) – Earth Science, (P) – Physical Science

PROFESSIONAL ARTICLES

There are many issues and changes confronting teaching professionals today, particularly in the area of science. In order to do the best possible job of meeting these demands, teachers need to have at their fingertips current information on the latest developments. It is for this reason that the following professional articles are presented in a spirit of cooperation and commitment to making you and your students' classroom experiences as meaningful as possible.

THE TEACHER'S ROLE IN FOSTERING POSITIVE ATTITUDES TOWARD SCIENCE

LaMoine Motz

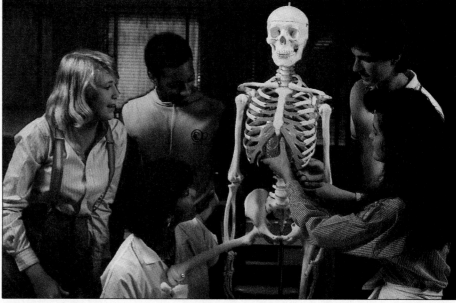

A number of years ago I observed a seventh-grade general science class in an urban school. I was told that the teacher, who turned out to be an ominous figure in a white lab coat, was "experienced" and "one of the best" they had. Throughout his lesson, however, such matters as lesson focus, objectives, and development were only hinted at. Were it not for the fact that these items were listed on the chalkboard for the students to copy, no one would have ever guessed what they were. At the end of the class, the teacher distributed worksheets to the students listing the vocabulary words for the week, which would be tested on Friday. When the period ended, the students filed out of the room.

When I recall this experience, I can't help but feel that the same scene has been played and probably still is being played in too many science classrooms thoughout the country. And it is disturbing to think how many more chalkboards will be filled tomorrow, and how many more vocabulary words will be defined on worksheets and tested on Friday. Is it any wonder that students do not like science? Is it any wonder that one of the main reasons students are bored with science is because of the teacher, who "makes us take notes," "goes too fast," "assigns large reading sections," and "makes use memorize terms." Science educators nationwide

have been discouraged by continuing evidence that most teenaged students dislike science, not as a subject in itself, but by the way it is generally taught in the school, and that many students underestimate the personal benefits of studying science. The task before us, if we are to help students reap the rewards of science literacy, is to foster positive science attitudes. The good news is that it can be done.

Attitude Development

The development of favorable or positive attitudes toward science depends on the curriculum and on the teacher's attitudes and practices in the classroom. As teachers, we often neglect the development and fostering of positive attitudes in science education instruction, focusing instead on content. Yet without these positive attitudes, which include motivation, interest, and a belief in the relevance of science to daily life, learning will not take place.

The key to building positive and realistic attitudes lies largely in the attitudes the teacher conveys and the atmosphere created for learning. Teachers need to radiate enthusiasm for science and show an eagerness to interact with students. Students should be given more open-ended, success-oriented experiences to help foster problem-solving, critical thinking, and decision-making skills. Many research studies in science education have confirmed the fact that students learn best when they experience science "in action" or via a "hands-on" method of experimentation.

Moreover, teachers need to ask themselves if they are projecting constructive, positive feelings about students. If a teacher has confidence in and visibly acknowledges a student's worth, that student feels better and works harder.

Even though teachers may have the best intentions, they sometimes project distorted images. In some cases, it is the personal distance the teacher maintains for authority that prevents students from responding to the subject matter. So while a teacher must be competent in the subject matter, he or she should be mindful of the role model students see at the front of the classroom, that HOW something is taught is important, if not more important, than WHAT is taught.

The attitudes formed in the classroom are changeable by a change of instructional attitudes, strategies, and techniques. That is where a good teacher can extend the values of the best textual material.

Good Science Teaching Makes a Difference

A recent study conducted at the University of Iowa on student attitudes has produced some encouraging findings. In the study, nine hundred middle school-aged students enrolled in four National Science Teachers Association (NSTA) Search for Excellence in Science Education (SESE) programs completed a questionnaire similar to ones completed by students taking the affective batteries of the National Assessment or Educational Progress (NAEP) assessments. A comparison of the results of these two studies reveals considerable differences between the attitudes of students in the NAEP schools versus science students enrolled in the NSTA exemplary programs. Many more students expressed positive attitudes towards science in the four schools with exemplary programs than in the NAEP randomly selected schools. A summary of the findings and comparisons shows that:

•Twice as many of the exemplary program students reported that their science programs were fun, and only half as many described their classes as boring.

•Almost 70% more of these students described their classes as interesting.

•Nearly three times as many exemplary program students reported that science classes enhanced their

curiosity, and twice as many reported that the classes prepared them to make decisions.

•Approximately 17% more of these students found science study useful in daily life, and more than 50% of the students reported that science study helped them to make choices.

•Students' opinions on the usefulness of science class as a preparation for further study were approximately the same as those of students in the randomly selected programs (both groups gave high ratings for such preparation).

What features of such exemplary programs produce positive student attitudes like these? Science programs that meet students' personal needs, help students to understand societal issues, enhance students' awareness of careers in science and technology, and are taught by enthusiastic, caring teachers are the types of programs that help students develop positive attitudes.

What common characteristics of teachers help students to develop positive attitudes towards science? The findings showed that:

•Far more students enrolled in exemplary programs felt that their teachers "welcomed" students' contri-

butions in the class than did students in the random sample. Twenty-five percent more of these students reported that their teachers like to hear students' questions, and nearly twice as many felt that their teachers wanted students to share their ideas in class.

•While both groups reported that their teachers generally knew their subject, 44% more of the students in exemplary programs agreed that their teachers knew a great deal of science.

•Students in exemplary programs reported 45% more often that their teachers made science exciting.

The evidence is in. A teacher can enrich, magnify, and give human values to the coldest facts, and can rise from being just a teacher to being an "educator."

Ignite The Desire To Learn

Children are curious and loving. Why can't we as science teachers nurture that curiosity and kindle the love? Children are also dynamic and energetic. Let's make our science that way, too. Whisper, shout, encourage, praise—be yourself, be real. It doesn't hurt. We light many matches in science. Perhaps the most important one is the one that ignites the fire for learning in our students. ∎

From Theory to Practice

Prentice Hall Science . . .

• Motivates students with a unique writing style, blended with dynamic visuals.

• Integrates relevant issues and day-to-day applications of science, emphasizing those that are of interest to middle/junior high school students.

• Provides numerous hands-on opportunities in a variety of formats — Discovery Activities, Laboratory Investigations, Doing, Discovering, and Activity Bank activities in the *Student Edition* and *Activity Book* plus *Prentice Hall Product-Testing Activities by Consumer Reports.*

• The Integrated Learning System offers a variety of learning vehicles, such as interactive videodiscs and videos, to keep student interest high.

There has been an explosion of scientific information and knowledge in recent years. Science teachers and curriculum developers have responded by trying to cover more and more science content in the same amount of class time. Yet current research on science teaching and learning provides clear evidence that this instructional strategy, which inundates learners with a plethora of facts, is failing our students.

Concurrent with this knowledge explosion is a growing consensus among science educators and researchers that "less is more"—that covering a few science topics in-depth rather than presenting many topics in a fleeting survey fashion has a more lasting, positive effect upon students. In short, students can learn more by learning less.

Why Less Is More

To most students, adequate coverage of a few topics is more satisfying than superficial coverage of many. Students in general tend to develop psychological ownership of problems and learn more when they study them over an extended period of time. Thus, "spacing and sequencing" facilitate optimal student learning. In contrast, superficial coverage often accomplishes little more than just familiarizing students with facts, terms, and definitions. Yet many students are currently memorizing facts about many different topics without understanding any of them in-depth.

Science education has too often proceeded as if scientific knowledge

In Science Today
"LESS IS MORE"

LaMoine Motz

consisted simply of lists of facts, concepts, and process skills, as if students could become scientifically literate by learning enough of the items of these lists. Unfortunately, most students who are exposed to science and learn science in this way do not become scientifically literate. In particular, most students fail to see the connections between what they study in science class and their personal understanding of the world. The disconnected facts and skills that they do learn are not useful or meaningful to them, and thus they are usually forgotten. Moreover, this concentration on committing facts to memory robs students of the opportunity to develop the thinking and problem-solving skills needed to deal with an increasingly complex world.

Scientific Literacy Is the Goal

The growing "less is better" consensus points the way to deeper coverage of fewer science topics. Such deeper coverage makes it possible for teachers to emphasize the process skills that will benefit students throughout their lives. Such skills include the ability to use, analyze, and construct scientific knowledge; to describe and explain real-world objects, systems, and events; to draw conclusions; to make predictions about future events; and to design systems and decide courses of action.

After developing such thinking tools, students can use these skills to understand the world around them and to guide their actions. They can develop solutions to problems that they encounter or questions they have. In developing solutions, such students can use their own knowledge and reasoning abilities as well as seek out additional knowledge from other sources. They also become learners as well as users and analyzers of knowledge, possessing the ability

teresting and motivating for the student.

4. You will be presenting how science works and the processes and methods that expand scientific knowledge and literacy.

5. Students will be the center of action, and you, the teacher, will be free to design the types of experiences that best fit your students.

"Less is more" is the glue that can make your science program exciting and enjoyable and your students more curious about the world in which they live. "Less is more" is the foundation and framework that can facilitate science to be a lifelong enjoyment and learning experience for the student. ■

to ask questions about the world that can be answered by using scientific knowing and techniques. In essence, "less is more" brings students closer to the ultimate goal of scientific literacy.

Not Just a Slogan

"Less is more" is not merely a slogan. It stands for an important belief: in becoming knowledgeable in science, as well as mathematics and technology, and probably in most other fields as well, the depth and applicability of understanding achieved is much more important than the quantity of material covered. Less "stuff" will be covered, fewer facts will be remembered for the test, and progress will seem, sometimes, to be exceedingly slow. It is a process of uncovering and discovering rather than just covering.

Where Do We Go From Here?

Recent studies on how people learn science have yielded some significant guidelines for the science classroom. Among the points of agreement are:

- Educators must take advantage of the efficacy of spaced learning.
- Teachers must provide students with experience in science phenomena before they are faced with terminology.
- Instruction must build on concepts of science in repeated experiences in different contexts.

- Learning must begin with direct experience and concept building.
- Skill in solving real-world problems requires extensive practice in solving real-world problems.
- Instruction must emphasize less material, but that material must be presented in more depth and with more meaning.

Applying "Less Is More" in the Classroom

If you are teaching from a "less is more" science curriculum, here is what you should observe happening in the classroom:

1. Students should be actively experiencing science, rather than just reading about it or passively hearing about it. In fact, there should be hands-on science at least two days a week.

2. Instruction should emphasize not the factual knowledge, but rather concepts that are developed and enhanced by making connections among the disciplines of science.

3. You should be covering less material, but making that material more in-

From Theory to Practice

Prentice Hall Science . . .

- Allows you to teach only the topics you want to teach and to teach them in-depth.

- Fosters critical thinking and problem solving with creative, real-life situations to think about and solve.

- Provides numerous activities, in a variety of formats, that are key to an in-depth understanding of the concepts.

- The multi-media components of the Integrated Learning System provide different pathways for developing each concept in tremendous depth.

INTEGRATING THE SCIENCES

Anthea Maton

Do you teach students the effects of water pollution in life science? Or is this topic covered in an Earth Science course as an extension of the material on the water cycle?

You could make a good case for including water pollution in either content area. Perhaps you might even include this topic along with other material in the chemistry components of a physical science curriculum. You certainly could include this material in the ecology component of a life science course. Maybe you could include water pollution in physical science when you study how heat produces convection currents.

If you have ever worked on a curriculum development committee, you know that problems similar to this one occur often. Sometimes topic placement involves only two disciplines, sometimes it may involve all three science subject areas normally taught at the middle school level.

Why Integrate the Sciences?

Have you ever wondered why life science is traditionally taught in the 7th grade, earth science in the 8th grade, and physical science in the 9th grade? Have you ever wondered why biology is taught before chemistry? Actually, when you think about it, this last case makes little sense, particularly when you need to understand some chemistry in order to understand biology. And in fact, most biology textbooks spend the first few chapters providing students with a brief chemistry course before the actual "biology" content is covered.

The reason often provided for placing physics in the eleventh or twelfth grade is that prior to this age students are not considered mature enough to deal with this subject matter. You may be surprised to learn that in many countries physics is taught from age 11, and in some countries physics is introduced at the very beginning of a student's school career. Of course the age at which physics is taught dictates the level of sophistication of its treatment. Whether you are studying running, a bouncing ball, or heat, it is all physics. Once the distinctions of grade level and subject matter have been determined by habit and convention, change becomes difficult—if not almost impossible.

How Can I Change the Teaching of Sciences Today?

If you think the sciences should be taught differently, you might want to begin to inaugurate a new science program in your own school and may be wondering how to go about it. Consider the following ways of presenting science. The explanations offered are not meant to be unchangeable. There is room for you to experiment in developing science curricula. Keep in mind that the terminology used is not always used the same way in every publication.

Semi-coordinated A semi-coordinated treatment would involve teaching the disciplines separately while holding "team" meetings so that each subject teacher is aware of what is being taught by teachers in the other disciplines and can refer to them when appropriate.

Coordinated In this approach, the four disciplines—physics, chemistry, earth science, and biology—are taught separately and in equal measure in the same class. However, students study an over-arching idea covered in each discipline concurrently. For example, light may be the over-arching idea being studied in all four disciplines in one semester. From a biological view point, the class might discuss the tendency of plants to grow toward light, and also the need for light to enable plants to undergo photosynthesis. From a chemistry viewpoint the class might study the different colors produced when elements are burned. They may also study the electromagnetic spectrum that could be applied later to spectral analysis. From a physical science viewpoint the class might study reflection and refraction followed by the physics of color. From an earth/space science viewpoint the class might study the electromagnetic radiations that emanate from stars and the colors of various stars seen from Earth.

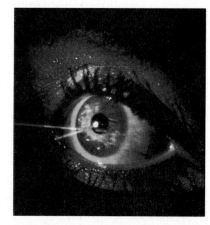

As you can see, in a fully integrated scheme, the students study a broad topic, in this case light, from the viewpoint of the four different disciplines. The students are able to study the unity and diversity among the disciplines as they study each science.

Integrated The following are two approaches for integrating the sciences:

- *The science is taught in modules.* Each module is based on one discipline, but draws from the other disciplines to various extents. Modules are designed so that a balanced science program is achieved over the course.

  For example, simple machines is mainly a physics topic, but you could also integrate material on the structure and function of human limbs. You could include material on the action and use of levers, inclined planes, gears, and pulleys. In addition, you might include applications of simple machines in the design of health club exercise equipment. Information on the use of simple machines used to construct Egyptian pyramids and Stonehenge could be included as well.

- *Merging the Disciplines* This technique of integration eliminates any discipline boundaries. Science subject matter is taught without regard to a particular discipline.

  For example, a study of music might include how sounds are produced and the way they travel. Examples include how sounds are heard by the human ear and differences in the ability of organisms to "hear" sounds. Topics covered could include the design of loudspeakers, devices to enhance hearing, and the design of concert halls to maximize hearing the sounds produced by orchestras, bands, and rock bands. As you can see, music is the single subject but is discussed through aspects that include several different disciplines. No single teacher is required to teach all aspects of music. When appropriate, for example, the biology of music will be taught by a biology teacher. The physics of sound will

be taught by a physics teacher. The teaching of the topic of music, however, is integrated throughout the different disciplines.

Other Ways of Integration

Some educators are currently discussing the integration of science with technology and society. For example, they feel that many environmental issues relate directly to certain aspects of science. They also feel that technology stems from an application of scientific knowledge as well as it being an important tool in scientific investigations. In this type of integration, the students play a major role in the direction the class takes.

Some educators also suggest building integrated curricula around environmental issues. This type of curriculum would be similar to a "science, technology, and society" curriculum in that the topic being studied would come first and the science taught would evolve from the topic.

It is important to remember that science is constantly changing as new information becomes known. The teaching of science changes in part as a response to the changing demands of an increasingly more complex society. A complete rethinking and restructuring of the way science is taught in American schools must occur within the next decade if the goals of a scientifically literate society are to be met. ■

From Theory to Practice

Prentice Hall Science . . .

- Frequently makes connection to other sciences in the content and visuals, as well as the Connections feature in every chapter.

- Assesses student knowledge as it applies to other areas of science with Section Review questions labeled Connections.

- Begins every chapter in the *Annotated Teacher's Edition* with an overview of how other sciences, as well as other curriculum areas, may be integrated into that chapter.

- Indicates with a blue number in the *Annotated Teacher's Edition* opportunities to integrate science and other disciplines, supported by teaching strategies on the same page.

- Allows you to select and configure the books to work with varying degrees of integration–from minimal to fully integrated.

The Thematic Teaching of Science

Anthea Maton

If you stop to think about it for a moment, organization is one of the basic features of life. Cells function in an organized way. The organization found in the universe is stunning. Even the parts of a car moving down the freeway move in an organized way. The human mind is organized and strives to organize the world around it. Educators are looking at ways to organize the teaching of science. One of the newest approaches is the teaching of science around themes, or larger topics. This type of integrated curriculum is a particularly suitable approach to teaching science at the middle school level, where efforts to coordinate the different subject areas around a single topic or theme are often used.

Outside of a classroom situation, students often attempt to organize material by using a thematic approach that actually tries to integrate materials from different science areas. As a teacher you become aware of these attempts at thematic organization if you listen carefully to the questions students pose during class. For example, in a discussion of the food-making ability of green plants, students may ask questions about light. They may wonder about the nutritional value in the materials produced by plants. A discussion of the gases used by plants during photosynthesis can lead to a discussion of the composition of the air, which can lead to a discussion of air pollution. You can see in this relatively simple example that science content in the mind of a youngster is far-ranging. In schools where science curricula are not taught in an integrated manner, the answer to some of the questions posed by students in a life science class can easily include topics normally covered in physics, chemistry, or earth science classes. This article will take the reader through the various stages involved in developing science curricula based on a thematic approach.

Selecting a Theme

It is important to keep students' interests in mind when deciding on themes in which science content is to be taught. In each stage in the development of a thematically arranged science curriculum, relevance to students' interests should be kept firmly in focus. A curriculum relevant to students' interests will also contribute to lively class discussions. Interest in a theme will lead to questions posed by students that will stimulate further discussion and insure the success of a well thought out, integrated curriculum. Some suitable themes are

Energy	Scale and Structure
Evolution	Systems and Interactions
Patterns of Change	Unity and Diversity
Stability	

The themes you choose must be capable of being expanded. Themes should offer opportunities for the inclusion of other areas of science in a "natural" way. Do not select themes in which the integration of other areas of science will appear forced. Themes that represent natural connections made by students will be most successful.

Expanding the Theme

Now the fun begins. Expanding the theme is best approached as a group exercise. Brainstorming enables the full potential and implications inherent in a theme to be explored. Brainstorming also removes the pressure a single person would feel if he or she had to think of possible interscience connections present in a theme. A group exercise will also give the teacher an idea of how exciting integrated science can be.

When you begin, examine the theme more closely. Write down the main aspects of the theme. For example:

Energy	*Systems and Interactions*	*Scale and Structure*
Light	Gravitational	Atoms
Earthquakes	force	Molecules
Sound	Water	Cells
Sun	Sound	Classification
Plants	Atmosphere	Ecosystems
Animals	Land	Tissues
Food	Machines	Organs
Water cycle	Chemicals	Organ systems
	Human health	
Unity and Diversity	The universe	
	The solar	
Organisms	system	
Genetics		
Matter		
Rocks		
Minerals		
Weather		
Natural resources		

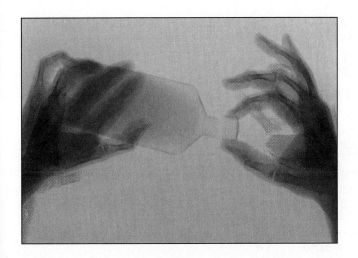

development by reviewing the themes included. A review will also help balance the content among the different scientific disciplines.

For example:

Scale and Structure

Physics and Chemistry

Atomic structure	Everything is made of atoms. Atoms are composed of smaller particles.
Universality of materials	Elements are materials made of the same kinds of atoms.

Biology

Cell structure	All living things are cellular. Cells are made of atoms. Cells contain smaller parts.
Plants and animals	Plant cells and animal cells have similar structures and unique structures.

Evolution	*Patterns of Change*	*Stability*
Environments		Density
Habitats	Weight	Environment
Genetics	Air pressure	Heredity
Fossils	Nutrient	Natural
Cycles	cycles	selection
Earth's crust	Classification	Atmosphere
Earth's weather	Genetic traits	Climate
Stars	Geologic	Energy from
Solar systems	history	the sun
	Oceans	Physical
		change
		Homeostasis

Completing the Unit

Now is the time to refine each unit. At this point it is important to consider the length of time that is to be allotted to each unit. Units as short as two weeks or as long as ten weeks may be appropriate.

Maintaining an emphasis on activity-based science should be uppermost in the mind of a curricula developer. Include in your planning: chances for projects, field-trips and other visits outside the classroom, guest speakers, process skills to be learned, and the availability of suitable audio-visual materials. Variety is the key to maintaining student interest in classroom work. Changing the teaching structure to a thematic approach does not mean you must change successful teaching techniques. ■

Further Theme Expansion

Now take each theme and expand it further. Consider the age of the students for which each unit or theme is intended. Different aspects of each theme and appropriate concepts can now be brought out by your brainstorming group. For example:

Scale and Structure

Atoms	Everything is made of atoms. Atoms are composed of smaller particles. Elements are materials made of the same kinds of atoms.
Cells	All living things are cellular. Cells are made of atoms. Cells contain smaller parts. Plant cells and animal cells have similar structures and unique structures.

Balancing the Unit

As you can see, each theme can be expanded. The expansion of a theme is limited only by the connections proposed and developed by the people discussing the theme. In the beginning, theme development may appear haphazard. However, you may refine the process of theme

From Theory to Practice

Prentice Hall Science . . .

- Uses seven common themes throughout the program–Energy, Evolution, Patterns of Change, Scale and Structure, Systems and Interactions, Unity and Diversity, and Stability.

- Provides thematic matrices in the front of each *Annotated Teacher's Edition* relating the themes and big ideas from each chapter.

- Precedes each chapter with four pages of teacher support that includes a thematic focus for each section.

- Allows you to select and configure the book to work with virtually any thematically based curriculum.

Building a
Better Science
Program Through

INTEGRATION ACROSS THE CURRICULUM

Dale Rosene

It is a rare individual who might try to build and equip a house on his or her own. Plans must be drawn, cement poured, bricks laid, woodwork cut and assembled, electrical and plumbing supplies installed, not to mention all the finishing touches that are involved. It takes training and skill to become a master mason, carpenter, electrician, plumber, or decorator. To become a master of all is nearly an impossibility for any one person. By bringing together a crew of experts, however, the finest of structures can be created.

Integration can take a number of forms. It can be as simple as having the math teacher review graphing before the science teacher starts a lab activity, or as complex as a carefully developed month-long unit involving a number of teachers. The best elementary teachers have been doing this very thing in their classrooms for years. Because they usually have the same students for all the subjects each day, they find they can make more efficient use of their time and be more effective in bringing about student learning by *blending* together the many subjects they teach each day. As an example, it is far more logical to have students read and write about the meal worms they are measuring for a science experiment than to teach students totally disconnected lessons in reading, writing, arithmetic, and science.

A commitment to understanding the needs of young people can be the first building block of this restructuring. Middle-level students are moving toward becoming abstract thinkers. As such, they learn better in an environment where they can see how things fit together, where they can make sense of ideas that span many subject areas, and where

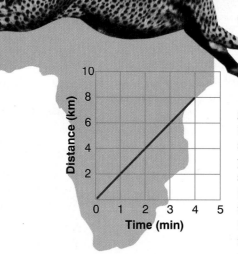

they can analyze and evaluate. Simply focusing on unrelated facts from a variety of classes doesn't allow them to use their newfound abilities. It is far easier for a teacher to be a "sage," lecturing each day and filling students' heads with every sort of important fact, than to be a "guide" who

provides a variety of coordinated experiences for his or her charges. Educators willing to make the extra commitment of time and energy will build a better understanding of the quickly forgotten facts they formerly dispensed.

Creativity and communication go hand in hand in adding to the materials necessary for this restructuring. Though a good teacher may see some of the connections between his subject and those of others, it is only when the experts begin talking about their curricula that the real opportunities for integration become apparent. Once the similarities are found, the participants must look for creative ways to tie them together.

As an example, a unit on scientists and inventors can integrate science (the scientific method, trial and error, and controlling variables), social studies (history and importance of significant inventions and discoveries), language arts (writing and research skills), math (measurement and the use of formulas), art (drawings of or advertisements for the various inventions), and even physical education (looking at the various new pieces of exercise equipment on the market). A student invention fair can be the culminating activity for the unit, with everyone creating his or her own real or fun invention to share with others. More adventuresome teams might build from less firmly connected topics, such as *Life in the Middle Ages*, and stretch their imaginations to find ways to teach similar ideas and skills from a totally different approach.

Like the house, a meaningful education can best be constructed by a team of experts working together, integrating the subjects they teach. Their materials are creativity, planning, flexibility, communication, commitment to students, and time. The experts are the classroom teachers, either as members of a teaching team or working together in a less formal structure. By coordinating their efforts, they can bring structure and meaning to their students' educational experiences.

Integration sometimes begins by chance. Content-trained teachers may "put blinders on" to what is happening in other classrooms in their school. It is only when the various curricula accidentally happen to mesh ("We learned about that in math class!") that a new awareness might appear. The math teacher is generally better prepared to teach youngsters about decimals than the science teacher who finds it necessary to review this topic before the density experiment. When teachers begin to talk about "teaching that together next year," the foundation is laid for looking for new ways to do things.

Flexibility is glue that holds the new structure together or allows it to fall by the wayside. Teaching together is often not easy. The best-planned units often will not follow the carefully structured time line set out before the unit begins. The teachers involved must be flexible enough to realize that they may have to adjust to these changes. They must be flexible in giving up some of the time from their science class to allow a special activity from another class to proceed. They might even have to adjust their teaching style to better fit the group activity. (It is difficult to be highly structured while students are enjoying a medieval feast.) Un-

less this flexibility is inherent in the planning of the cooperative effort, communications will break down and the integration will suffer.

Adequate time and planning are the final essentials for integrating science with the rest of the curriculum. Time to plan, time to gather and create materials, time to teach and learn, and time to evaluate what was accomplished are all necessary. While planning time is ordinarily built into the teacher's day, some planning may have to take place outside of school hours. Major units could be planned several months in advance, while smaller teacher-to-teacher efforts might be discussed one day and carried out the next. To be successful, however, it is absolutely necessary that all teachers involved know when activities are taking place and what their commitments to the project are. As such, face-to-face planning is essential. Adequate time to plan and teach are equally important.

To make the building of a house more efficient, a contractor coordinates the efforts of the skilled tradespeople. The contractor obtains the necessary materials and permits, schedules the various skilled workers as they are needed, and works to ensure that all jobs are done properly. An integrated curriculum needs some kind of contractor as well. This job can be done by different individuals for each cooperative effort, but it is critical that one person coordinate efforts, whether for two teachers working together or for a whole team of teachers. This person will maintain the communication necessary to keep participants flexible and coordinate all aspects of the planning necessary to make the project successful.

Just as every house has a door and windows, our classrooms are similarly equipped. While it is all well and good to look through those windows to see how science is related to other subjects, it is more important to open our classroom doors and our curriculum and invite the other subjects inside. They make good company and we'll want to invite them back often.

From Theory to Practice

Prentice Hall Science . . .

- Frequently makes connections to other curriculum areas in the content and visuals, as well as in the Connections features in every chapter.

- Consistently provides opportunities for connecting science to literature and mathematics with Reading and Calculating Activities.

- Assesses student knowledge as it relates to everyday life situations with Section Review questions labeled Connections–You and Your World.

- Alerts teachers to any cross-curricular tie-ins in the student text with a blue number in the *Annotated Teacher's Edition*. The teaching strategy at the bottom of the page follows up in more depth with suggestions for integrating the cross-curricular connections.

- Provides Integration notes in the *Annotated Teacher's Edition* margins for additional high-interest tie-ins to other subject areas.

VARIETY:
THE SPICE OF SCIENCE

Anthea Maton

What are some factors that make a good science lesson? How important is the teacher, the students, the subject matter? What part does the school play? Obviously, there is no single factor that makes a lesson successful. However several components that make up a good lesson can be discussed. This article will concentrate on several factors that add quality to science teaching.

Today many students consider science classes to be difficult and boring. In fact, students who early in their school careers state that science is their favorite subject, often shun science in secondary school. What happened in the intervening years to turn these students off? How can we prepare students for life in a scientifically complex world if they avoid science as if it were a serious disease, that once caught could lead to terminal boredom? It is obvious that we must quickly do something to rekindle an enthusiasm in all students for science.

The idea of science for all students is embodied in "Project 2061: Science for All Americans" from the American Association for the Advancement of Science, and "Project, Scope, Sequence and Coordination" from the National Science Teachers Association. While a discussion of these projects does not fall within the subject area of this article, it is important to note that both projects involve a greater involvement of the student in "doing" science.

The trend toward a lecturing and questioning style of teaching needs to be changed. Science classes need to be enriched through more laboratory investigations and activities. The study of science needs to be expanded beyond the walls of the classroom. Resources sometimes limit laboratory work. However, there is evidence suggesting that the teacher is the most important factor in interesting students in studying science. Many successful science classes can be observed in areas with limited resources.

The classroom teacher is responsible for creating a positive classroom attitude. A teacher should encourage ways of examining the natural world and "involve" his or her students in science. Through attitude, a teacher can exert a profound influence on students. To be successful in science, students must be introduced to new ways of looking at things. Mary Budd Rowe has researched various aspects of science education. She suggests that students should be encouraged to ask a series of questions. Ms. Rowe groups the questions into four categories:

Ways of knowing
  What do I know?
  Why do I believe?
  What is the evidence?

Actions/Applications
  What do I infer?
  What must I do with what I know?
  What are the options?
  Do I know how to take action?
  Do I know when to take action?

Consequences
  Do I know what would happen?

Values
  Do I care?
  Do I value the outcome?
  Who cares?

It is through the use of these and other similar questions that Ms. Rowe believes students start to wonder about the world around them, about the role they play in it, and about the influence they may have on it.

Ideally, activities that enable a student to experience phenomena firsthand should come at the introduction of a new topic. In this way, a sense of discovery is built into the topic and the students are able to build a knowledge of concepts from their own experience. Students should no longer be doing "experiments" to verify others' results or to mirror a demonstration. They should be doing experiments to discover things for themselves. Teachers should structure lessons to incorporate the exchange of ideas among students. The communication of ideas to others, and research into other people's experiences are a natural part of "doing" science. It has been suggested that a teacher's role should be much like a Principle Investigator in a research laboratory, with the students being the scientists. The students are then given opportunities to carry out experiments on suggested topics, communicate their findings to colleagues, compare their results with others, have access to scientific journals and magazines, and even to present their ideas to a peer-group audience.

When advocating a similar approach to the way work is done in a "real" laboratory, it is important to include the debate on the so-called scientific method. Much has been written about the scientific method and the various stages that occur during an experiment. Two issues arise. First, the scientific method, in some ways, can intrude on the actual carrying out of an experiment. Students sometimes stick too closely to a rigid format—a format that often becomes more important than the actual investigation. Second, many of science's greatest discoveries did not arise from the accepted route produced by a rigid application of the scientific method at all. These discoveries often arose from a serendipitous happening. Students should be encouraged to view science, not only as a carefully planned and orchestrated affair, but rather as a world of chance discoveries, of months of work that often leads to a negative result, of sometimes years of painstaking observations, and of occasional flashes of insight or genius that can change the ways we view the world.

The hands-on approach is not the only discovery tool that the teacher has to offer students. A balance of different teaching methods involves students in all aspects of study. The use of some or all of the following methods will help make science accessible to all students.

Teaching Methods

Textbook work	Brainstorming
Laboratory work	(concept mapping)
Teacher and student	Role playing
demonstrations	Problem solving
Group work—	Questioning
cooperative learning	Journal keeping
Projects	Classroom discussions
Integrated technology and	
media (video/video disc)	
Independent study	

Some of these instructional methods are more appropriate at different stages of a course unit. A good teacher uses a variety of methods. Not included in this list are special resources, such as outside speakers, field trips, and home assignments. All add new dimensions to a science topic, broadening its scope.

Whatever methods are used, there is little doubt that all schools must take it upon themselves to provide a science education for all students. The rapid growth in technology has affected the lives of everyone. Science is everywhere, and it is up to science educators to make students feel comfortable with it. ■

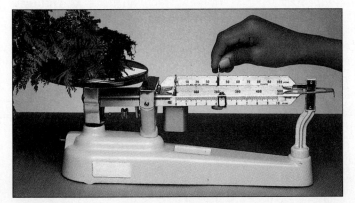

From Theory to Practice

Prentice Hall Science . . .

• Begins each book with a Discovery Activity where students can explore what they know about the topics presented.

• Provides various activities throughout the chapters designated as Doing, Discovering, Thinking, Writing, Reading, Calculating, and Activity Bank.

• Includes an in-text laboratory investigation in every chapter plus one to two additional investigations per chapter in the *Laboratory Manual.*

• Encourages students to think creatively as they solve the real-life situations presented in the Problem-Solving features.

• Includes an additional Discovery Activity plus two to three additional hands-on experiences in the *Activity Book.*

• Provides real-life application of science knowledge and skills with *Prentice Hall Product-Testing Activities by Consumer Reports.*

• The multi-media components of the Integrated Learning System provide a multitude of possibilities for discovery learning and inquiry.

Problem Solving in the Science Classroom

Dr. George Ladd

A middle school teacher shows her class two beakers of equal size, each filled with the same amount of a clear liquid. The teacher asks whether an ice cube will sink or float when placed in the first beaker. When the cube floats, cheers break out from the students who answered correctly. The teacher then asks the students what they think will happen if the same ice cube were placed in the second beaker. Most respond that the cube will float. A collective gasp can be heard when the cube sinks! A problem now confronts the class, and with it an opportunity for problem solving.

What does it take for a teacher to create a problem-solving environment? The driving force behind the act of problem solving is curiosity, an interest in finding out. In the 1930s, John Dewey stressed that the problems that confront students must be perceived by learners to be real and interesting in order for their curiosity to be stimulated. Therefore, if teachers want to involve students in problem-solving activities, the nature of the problems selected are crucial.

Teachers want their students to learn. What the teacher teaches can be cognitive (information or knowledge), affective (attitudes and values), and/or psychomotor (manipulative). Regardless of how it is classified, all knowledge has but one origin: the process of answering questions. The asking and answering of questions is what problem solving is all about.

The Four Steps of Problem Solving

While theoreticians disagree on the exact number of steps a learner must take to solve a problem, there are four basic steps that are essential to the problem-solving process.

1. Identifying a problem and describing it in a way that can be understood
2. Determining what the outcome of solving the problem might be
3. Exploring possible solutions and applying them to the problem
4. Evaluating the possible solutions and revising solutions if they do not pass the test in other similar situations

These behaviors are commonly called critical thinking. Without these behaviors, the solving of problems presented by the natural world around us cannot occur. Additionally, such qualities as curiosity, open-mindedness, suspended judgment, and tolerance for ambiguity, and the like must be present. Moreover, in solving problems, pupils need to use knowledge from their past experience, transfer it, and apply it to the new conditions confronting them.

Several key educators have recently helped to operationalize the problem-solving activity by identifying it as a student behavior and not a teaching strategy. To wit, a science

teacher can create an environment in which problem solving can occur, but the teacher cannot carry out the behavior for the child.

Problem Solving Versus Discovery and Inquiry

Literature focusing on problem solving in science often contains terms such as critical thinking, discovery, and inquiry. Some authors use some of this terminology interchangeably, adding to a growing confusion among science teachers as to what problem solving actually is and how it impacts their everyday teaching.

What about discovery and inquiry? Where do they fit in as critical thinking and problem-solving behaviors? You may have read and heard that discovery and/or inquiry learning both contribute to critical thinking and eventually to the solution of problems. This is simply not the case. Discovery and inquiry are both instructional strategies that provide settings for critical thinking and problem-solving. They are NOT students' behaviors at all!

The difference between discovery and inquiry rests in the origin of a problem and who controls learner behavior in problem solving. In the discovery strategy, the teacher (or some other source external to the learner) identifies both the problem and the process a learner goes through to solve the problem. The only student control comes with the opportunity to identify a tentative solution. For these reasons, discovery teaching is often preceded by the term "guided" to further emphasize the external control present in the teaching strategy.

Inquiry, unlike discovery, is not a linear process. Inquiry can be promoted by discovery, but it is not the same. Inquiry may begin with a teacher-selected problem, but from that point on, the control moves to the learner. The learner decides how to carry out problem-solving processes and how to identify solutions without interference from any outside source. This condition dictates that the learner practice higher-ordered thinking skills. Adolescents, who are developing the ability to carry out formal thought, need this practice if they are to be able to solve the more abstract problems they will eventually face.

Both discovery and inquiry certainly have their place as part of problem solving. Each provides opportunities for critical thinking (whether teacher or student selected). Each remains important in gaining a better understanding of the world. Providing opportunities for pupils to engage in discovery and inquiry provides a challenge for teachers who use problem solving as part of science instruction.

Ensuring Success in Your Classroom

For problem solving to occur, first and foremost student curiosity must be stimulated with problems that engage, intrigue, and motivate. Students need to understand and know how to apply the four steps of the problem-solving process. They must be able to make connections with their past experiences, while keeping an open mind. Working as a facilitator, the teacher can help develop this through both discovery and inquiry teaching strategies. When these conditions are present, the result is a classroom rich with problem-solving possibilities and opportunities, a classroom where enthusiasm reigns, students learn, and lifelong problem-solving skills are cultivated. ∎

From Theory to Practice

Prentice Hall Science . . .

• Provides numerous settings for critical thinking and problem solving with Discovery Activities and Hands-On activities throughout the texts.

• Encourages creative and critical-thinking with high interest Problem-Solving features in virtually every chapter.

• Includes Critical-Thinking and Problem-Solving questions in every chapter review.

• Gives more opportunities for problem solving in the *Activity Book*.

• Assesses thinking and problem-solving skills with Performance-Based Tests in the *Test Book*.

• Applies students' problem-solving abilities to everyday use on consumer products with *Prentice Hall Product-Testing Activities by Consumer Reports*.

COOPERATIVE LEARNING and SCIENCE— The Perfect Match

Robert M. Jones

In cooperative learning, small heterogeneous groups of students work together to achieve individual and group objectives. The objectives may include both content and process skill outcomes. Additionally, by using well-defined roles under the supervision of the teacher, students can develop social interaction skills.

The research on cooperative learning is clear: It produces positive effects on students who participate in it. Cooperative learning methods contribute to improved student achievement at all grade levels and in all subject areas. In the affective domain, improved intergroup relationships among gender and ethnic groups and between normal and exceptional students result. Students' attitudes toward school improve. Recently researchers demonstrated that by applying the cooperative approach to the test review process, most students could achieve passing scores on chapter and unit tests. Low-achieving or "at-risk" students show increased learning improvement, with gains significantly higher than average- or high-ability students.

The good news for science teachers is that the cooperative learning techniques that have produced these dramatic results are ideally suited for application in the science classroom. While cooperative grouping has been used successfully with hands-on laboratory activities for many years, current studies have also validated the use of cooperative strategies for textbook-based content learning in science class as well as in the test-review process.

With these three different applications, today's science teachers have a wide array of options for implementing cooperative learning. Moreover, each type of cooperative grouping can stand alone or be used in combination with the others. Teachers may decide to select only one use of cooperative groups or they can combine two or all three.

The following sections provide science teachers with specific guidelines for utilizing cooperative learning in the laboratory, for content learning, and for test review.

> *Cooperative learning methods contribute to improved student achievement at all grade levels and in all subject areas.*

Cooperative Groups in the Laboratory

The key to achieving positive results in cooperative laboratory activities is to divide the class into small heterogeneous groups, to develop specific tasks for each student in the group to perform, and to integrate these tasks into a teaching/learning sequence. For classes structured into groups of four students, for example, the teacher would assign each group member one the of the following roles: Principal Investigator, Materials Manager, Recorder/ Reporter, or Maintenance Director. Roles can be combined or divided as group size dictates. Two special roles, Technical Advisor and Observer, can also be added. Role responsibilities are as follows:

Principal Investigator directs all operations associated with the group activity, including checking the assignment, giving instructions to the group, asking questions of the teacher, providing assistance to group members, and conducting group discussions about results. The principal investigator either directs the activity or assigns the conduct of the activity to other group members.

Materials Manager obtains and dispenses all materials and equipment for the laboratory activity. This student also sets up and operates the equipment in cooperation with the Principal Investigator and assists the Recorder/Reporter in collecting data. The Materials Manager is the only student who moves around the classroom without special permission during the activity.

Recorder/Reporter collects information, certifies results, and records them on the group worksheets given to the teacher. This student is also responsible for reporting the group results to the class. The results can be in the format of an oral report given from his or her seat or in a written report placed on a Class Summary Chart on the chalkboard.

Maintenance Director is responsible for cleaning up the work station and has the authority to assign other members of the group to assist. When all materials and equipment have been collected, this student directs the disposal of used materials and turns over activity equipment to the Materials Manager for return to the teacher's station. The Maintenance Director is also in charge of group and individual safety.

Technical Advisor is a special role assigned to students who complete work promptly and accurately. Technical Advisors are allowed to move to a group having difficulty with an investigation and to give assistance. They also carry out special assignments made up by the teacher, such as preparing materials for another activity or inspecting equipment and materials.

Observer serves as an administrative assistant to the teacher. Observers may use a form to record problems groups are having with their procedures, then return this information to the teacher who uses it to improve the instructional program. Each student should have the opportunity to serve as an Observer during the year.

The Teaching/Learning Model

The following six-step instructional model provides teachers with a format for quickly and effectively implementing cooperative laboratory activities:

1. *Introduction* Introduce the process skill science lesson to the whole class at one time. Before small group work begins, write the lesson objective and any pertinent activity information, such as vocabulary and safety requirements, on the chalkboard. Ask for and answer all student questions about the activity at this time.
2. *Distribution of materials* Each group's Materials Manager collects the equipment and materials from a central station and carries it to the group work station. (In some instances, where safety or cleanliness is involved, the teacher may choose to distribute certain materials directly to the group activity stations.)

Pam
Principal
Investigator

José
Materials
Manager

Vickey
Recorder/
Reporter

Ian
Maintenance
Director

3. *Group Investigation* After the materials and equipment have been distributed, the Principal Investigator goes over the directions to make sure everyone in the group understands the activity. The Materials Manager then directs the setup of materials. During the setup and investigation, the teacher moves from group to group observing the use of equipment, procedures, and results. If the group cannot resolve questions about the activity, the Principal Investigator may raise his or her hand and ask the teacher for assistance. The Recorder/Reporter collects all required data on a data sheet and, after certifying the results with other team members, posts the group results.

4. *Discussion of Results* After the data from each group has been posted, the teacher conducts a discussion of the results. During this important step, concepts and skills are clarified, similarities and discrepancies are noted, and, if necessary, certain procedures are repeated so that each student has a clear understanding of both the process skills and the concepts involved in the laboratory activity. The teacher can follow the discussion with reading assignments, handouts, and related activities.

5. *Group Station Cleanup* The Maintenance Director takes charge of the group and makes sure that the work station is clean and that all materials and supplies are returned to the designated area. The Materials Manager inspects the activity materials and returns them to the materials station. (With very young students or certain materials, the teacher may wish to order group station cleanup before the discussion of results.)

6. *Student Evaluation* Each lesson should involve the students in demonstrating mastery of the process skill objective. This often can be demonstrated and documented with the group data sheet. In other cases, pencil-and-paper test items can be used to evaluate concept and skill attainment. The teacher can also jot down notes on each student's role performance or design a form for student self-evaluation.

Hints and Suggestions

- Have students wear badges that readily indentify their specific role in the group. Badge designs can range from fanciful shapes and colors to very formal designs. Students can even design their own badges. For younger children, the badge can be hole-punched and worn around the neck.

- Let each group choose a name. This produces an identity association for students and a method of group identification for the teacher. Group names can be printed on the badges.

- Change group membership every three to six weeks. Study the sociology of the class and arrange groups of students with different ability levels, ethnic backgrounds, and gender. This rearrangement benefits students both cognitively and socially.

- In establishing cooperative laboratory group procedures,

talk only with the Principal Investigator. This will establish a chain of communication in the group and prevent a repetition of questions.

- To ensure clear communication, write and post class rules, group names, job descriptions, and other important information on the bulletin board in the classroom.
- Post job assignments and their rotation schedule in the classroom. This shows what activities will occur in the upcoming sessions and demonstrates that roles are assigned in a fair manner. Students should rotate jobs after each activity is completed.
- Let the roles, badges, and job descriptions work as a discipline system. Most disputes over task behavior can be easily resolved by referring to the job description or assignment sheet.
- Use the groups both indoors and outdoors. Many activities involving temperature, direction, measurement, texture, living organisms, weather, size, and shape may be conducted outdoors. For outdoor activities, arm bands or headbands work better than badges for identification.

Cooperative Content Study Groups in the Classroom

In addition to laboratory work, cooperative groups can be used effectively with the science textbook and other curriculum materials. In this application, mixed ability base groups or "Home Teams" composed of three to four students work together for several weeks using publisher-furnished or teacher-made materials to learn the facts, concepts, and skills in the chapter or instructional unit. Each group member must contribute to the group results and is rewarded for his or her contribution.

This cooperative content study group approach helps motivate students because everyone is involved with discussing and learning material. Individual student reading is minimized and student interaction is maximized through discussion and graphic techniques. The problems associated with "reading the chapter" as a precondition to participating in science are eliminated by using Home Teams for short, structured reading tasks and concept learning. The Home Team provides academic and social support for each member and team membership gives status to slow or disinterested learners who seldom receive classroom recognition. Moreover, the science curriculum can be taught more effectively because less classroom time is needed to teach the same material at greater depth.

To implement cooperative content study groups using Home Teams, the teacher should assign one high-ability, one low-ability, and one to two average ability students to each Home Team. Once the teams are formed, the following instructional steps can be used:

1. *The teacher introduces the unit to the class.* The unit overview may include a film, videotape, or other media to generate excitement about the topic. This introduction should also highlight the major divisions in the unit.

2. *Teacher distributes Home Team worksheets and makes assignments.* Students do not read the chapter before the Home Team meets. The questions and statements on the Home Team worksheets are used for reading the specific paragraphs in sequence.

3. *Students meet in Home Teams and learn concepts identified on the worksheet.* Students use the worksheets to read paragraphs, discuss concepts, and agree on responses. During this study session, the teacher moves among the groups and facilitates their work. A grade can be given to each student for completing the worksheet. Supplemental curriculum materials also can be used during this session.

4. *Teacher conducts whole-class summary of the worksheet.* This discussion session helps clarify content questions and eliminates the possibility that an incorrect response has been given in the Home Teams. At this time, the teacher can correct errors, expand concepts, and add explanatory information. Short demonstrations using manipulative materials and equipment can be conducted during this summary session. (Repeat steps 2, 3, and 4 for each Home Team worksheet. A grade can be given for completing each worksheet.)

5. *Students conduct one or more of the laboratory activities presented in the textbook or selected from some other source.* The Home Team can function as a laboratory activity group during this class session. Films, guest speakers, discussions, reports, and other extending activities can be included at this time.

6. *Home Teams use study items to review for the test.* After instruction is complete, Home Teams reconvene to prepare for the test. This session is important for slow learners and low-ability readers. The students should locate the study item responses and discuss them. One student should be in charge of this process at each group station. A grade can be given to each student for completing the study items.

7. *Teacher conducts whole-class review.* The Home Team can play games or participate in group competitions based on the test-review items. The classroom atmosphere should be positive and encourage participation.

8. *Students take individual tests and receive individual grades.* Up to this point, the Home Team members have received identical grades for the group work products. The individual unit test is used to assess student mastery of the concepts. The Home Teams are given recognition based on group improvement and sustained performance. Each student receives a unit score based on a weighted combination of the daily group work and the individual test score.

An interesting variation of the Home Team cooperative learning model is the "jigsaw" approach in which each member of the Home Team is designated an "expert" on one section of the chapter or unit of study. The experts for each section of the unit from all of the vari-

ous Home Teams meet in expert groups, which read, discuss, and learn the important material in their section. The experts then return to their Home Teams to share their expertise with their team members. The experts teach in turn until all the material in the chapter or unit is covered. This procedure, while requiring more social skills and movement in the classroom, is very effective and requires that each expert do his/her job for the success of the team.

Hints and Suggestions

1. Include worksheets on the processes presented in the chapter. Have at least one of these higher-level learning activities to accompany each Home Team worksheet. Use higher-level questions on the worksheets to ensure that the team members think about the material.
2. To keep teams on task, establish reasonable deadlines for Home Team activities. Have extra materials or meaningful tasks available for groups that finish early.
3. Pair a marginal student with an average or bright student for peer instruction and support (this may mean that some Home Teams will have extra members). This close support will help marginal learners improve their learning and leadership skills. Give "good citizen" awards to peer tutors.

4. Send the team worksheets home with a note to parents. Explain that these materials will help students improve their test scores. These sheets provide parents with a useful mechanism for working with their children.
5. To build team spirit and cohesiveness, encourage the Home Teams to participate in some fun activities. Nonacademic games and contests can include paper airplane and kite flying contests, construction tasks, and science trivia games.
6. Have the Home Team be responsible for updating members after absences; for maintaining and distributing pencils, paper, and other materials; and for routine maintenance and clerical tasks. This will help create a sense of shared team responsibility.

Cooperative Test Review Groups

Several years ago researchers at the University of Houston-Clear Lake discovered that by participating in cooperative test review teams, lower- and average-achieving students improve their test scores significantly and high-achieving students maintain their achievement levels, while developing a sense of social responsibility and desirable leadership skills.

In the cooperative group test review technique developed at the university, students are grouped in mixed-ability review teams composed of three to five members, with one high-achieving student, two to three average students, and one below-average student. The teacher furnishes each group with test-review materials that focus on the specific concepts and skills that will be assessed in a chapter or unit test.

With the guidance of an appointed group leader, each review team then advocates responses to the study items, clarifies concepts and skills through discussion, and makes sure each team member can correctly respond to specific items. The primary goal of the test review team, of course, is to ensure that each member has mastered the knowledge and skills represented by the study items, and it is understood that *each* team member will contribute to the test review process. The teams should remain intact for at least one grading period.

To implement this cooperative test review procedure in the classroom, the teacher should follow these steps:

1. *Locate or construct a test for the teaching unit.* Test items should be substantive and understandable and written in standard format.

2. *Develop one or more review items for each test item.* Review items should stress the specific knowledge and skills needed to correctly respond to corresponding test items. For vocabulary, facts, and concepts at the knowledge level, there should be one study item for each test item. At application and higher levels, two or more study items may be necessary for each test item.

3. *Assign students to mixed-ability test review teams, once all instructional activities have been completed.* If Home Teams are already established for academic tasks, they can simply reconvene as test review teams. If not, divide the class into mixed-ability groups of three-to-five members. At least one student per team should be able to assume a leadership role.

4. *Distribute focused study items to each review team and give directions for the review session.* Each student in the test review team will need a copy of the study items. Use oral directions or printed instructions to establish the tasks for each test review team and appoint one student in each group to be in charge of the session and to monitor team work. The role of team leader should be rotated at each review session.

5. *Move from group to group, answering questions and providing guidance as needed.* Because peer interaction, elaboration of facts and concepts, verbal and graphic expla-

nations, and group discussions are all important, it is critical to allow teams to work without undue interference. At the same time, the teacher should be available to provide assistance in case the review process falters.

6. *Conduct a whole-class review of study item material.* When test review teams have completed all of their tasks, each student should have a basic understanding of the concepts and skills that are important in the instructional unit. The teacher can utilize this whole-class review session to summarize, answer questions, and restate the key concepts and skills identified in the study items. To add an element of fun and team competition to the review process, teachers can devote all or part of this session to review games such as "Jeopardy," or "Family Feud."

This cooperative test review procedure is independent of instructional techniques and can be used in any classroom setting. Moreover, review teams may be formed *after* formal instruction is completed, giving teachers the option to use cooperating learning throughout the instructional unit or only for the test review process. Best of all, when students realize they can improve their test scores and report card grades, they actively cooperate in completing test review team tasks. ■

From Theory to Practice

Prentice Hall Science . . .

• Begins every chapter with a Science, Technology, and Society/Cooperative Learning activity suggested in the *Annotated Teacher's Edition*.

• Provides many of the Discovering, Doing, and Activity Bank activities in a format that can be used with cooperative groups.

• Laboratory Investigations in the *Student Edition* and *Laboratory Manuals* are designed to be used with teams of students.

• Includes activities designated as Cooperative Learning in the *Activity Book*.

• Fosters student teamwork in testing, as well as designing tests for, consumer products with *Prentice Hall Product-Testing Activities by Consumer Reports*.

Today movements to develop and implement national standards for science education are in various stages of development. Presently several states have enacted various statewide testing programs. Some issues currently under discussion include: Why do we assess student progress? How do we currently assess student understanding? What are the best ways to assess student understanding?

AUTHENTIC ASSESSMENT:
What Test? How to Test? These are the Questions!

Anthea Maton

Why Do We Need to Assess Student Progress?

It is important to begin with the assumption that all students are able to learn the material being presented. This is not to say that all students learn in the same ways, or at the same pace. Students differ in their ability to reach the same level of achievement on tests. Various learning outcomes also will result when different teachers emphasize different aspects of the subject matter. Therefore, a variety of testing methods need to be used to form a complete picture of the students' knowledge and progress.

The impact of assessment on students can be enormous. Tests often punctuate a series of periods of accountability. Achievement, or lack of achievement, on certain standardized tests are often responsible for "identifying" a student's intellectual capabilities. As a result of these tests, students are often placed in a particular "track." Ultimately tests may exert a profound influence over college admission and may even affect a student's career opportunities.

There is little doubt that it is a difficult task to design valid methods to monitor a student's progress in the use of skills fundamental to science. Assessment has to be integrated into the teaching plan and should be viewed as an important part of the teaching process.

What Do We Mean by Assessment?

Assessment is often viewed by students and teachers alike with feelings of apprehension. Yet assessment is of basic importance in the learning and teaching of science. "Learning" science involves an understanding of many concepts, the acquiring of factual information, and the development of many differing skills and processes. Most educators would agree that assessment means the monitoring of a student's knowledge, capabilities, progress, and understanding. Assessment can be viewed solely as testing, or it can be used to examine the effectiveness of the teaching taking place. Assessment should be helpful to both students and teachers. Because of its importance, the assessment process should be carefully thought out.

What Methods of Assessment Are Available?
How Useful Are They?

- By far the most heavily used assessment method is the *multiple choice* (MC) test. Multiple choice tests are easy to administer and easy to grade. Because of these advantages, multiple choice assessment has achieved a high degree of popularity among teachers. This type of test assesses recall. Multiple choice tests are primarily used to test a student's factual knowledge rather than a student's understanding of a topic. There are, in fact, some extremely well-designed multiple choice test banks in use, but many are very poor. Although multiple choice tests are not ideal for testing understanding, they often can be successfully adapted to this use. For example, by asking students to give a reason for selecting one answer over another, a student's understanding of a topic can be assessed.

- *Short answer questions* are another widely used method of assessment. Short answer tests range in complexity from simple fill-in-the-blank tests, to more complex tests that involve writing one or two sentence answers. The more students are able to write, the more useful these tests become.

- *Long answer or essay questions* allow students to probe more deeply into their understanding of the subject matter. This method of assessment also permits the teacher to explore a range of topics of varying levels of difficulty. In this method of assessment, a teacher has to be careful that students do not stray far from the topic and that the teacher does not give credit for writing style at the expense of science understanding.

- *Laboratory reports* are another fairly traditional method of assessment. This assessment method tests the students' ability to record and handle data. It also tests the students' ability to analyze, extrapolate, synthesize, and evaluate their findings in a logical fashion. Students' ability to communicate their findings to others is also assessed.

- The *oral report* is similar to essay questions. The students benefit from researching a topic. Students must also understand a topic well enough to be able to clearly, and logically explain it to others. Other students also benefit from this method of assessment as knowledge about different topics is communicated and shared.

• *Comprehension exercises* are often used to assess progress in Langauge Arts. This is a good and slightly more unusual assessment tool in science. The student is provided with a scientific article that contains material relevant to the topic being studied. Then the student is asked to write answers to questions on the article. Ideally, the answers come from the students' understanding of the topic or related topics.

Most of the assessment methods mentioned so far are used presently to some extent by every teacher. The assessment methods that follow may be new to some teachers, although these methods quickly are gaining importance in assessment circles.

• Rather than breaking the assessment of a student's work into fragments, the *portfolio assessment* method allows the student to present his or her work as a whole. While this assessment method is new to science, it has been used successfully in art classes for many years. There are no hard and fast rules as to the contents of a student's portfolio. For example, the portfolio can contain ten pieces of work that received the best grades, or the portfolio may include ten pieces of work that the student feels best represents his or her abilities. The portfolio can also contain all or a wide selection of a student's work. The teacher is encouraged to have input in choosing the method selected. It does not matter how the portfolio is assembled, providing both student and teacher are satis-

fied that the portfolio offers a true representation of the student's understanding and abilities.

• *Performance-based assessments* are a relatively new method to assess students' performance. Performance-based tests are a way to evaluate both process skills and students' understanding. If they are properly constructed, performance-based tests can also be an enjoyable learning experience for students.

Often in a performance-based test relatively simple tasks are required of students. These tasks often take the form of a simple laboratory investigation or activity. The tasks are, where possible, placed in an everyday context so that the student can view the task as relevant. By using unusual situations, the tasks are also designed to teach as well as to assess. Performance-based assessments are an exciting addition to the evaluation repertoire of a teacher. For many students, they also provide a more interesting alternative to other assessment methods.

What Is the Future of Assessment Methods?

Obviously technology will play a larger role in the classrooms of the future. Today several projects are underway that will use the interactive capabilities of laser-disk players and compact-disk players to enhance assessment methods. One project, from

the National Science Teachers Association uses interactive compact disks combined with performance based assessment to evaluate students, and to track the development of their thinking processes as they work through each assessment item. The information gained will allow teachers to work more closely with each student to improve weak areas. These assessment methods are being designed to apply to students of varying ability levels.

Whatever assessment method is used, it is important to remember that teaching and assessment go hand in hand. Assessment methods evaluate both teaching and learning. No single assessment method should be used exclusively in a course. Each method plays an important role in an evaluation scheme. It must always be remembered that a teacher's assessment of a student can exert a strong impact on student interest in a subject, and more importantly in students' self-esteem. During an assessment process, a few words of praise from a teacher can make students feel that they are an important part of the class, and are special to the teacher. ■

INTEGRATING TECHNOLOGY IN THE SCIENCE CLASSROOM

Amy Wagner and Richard Thibeault

The need for people who can design, maintain, and use the tools of the technological age is evident. These tools are everywhere—from automatic teller machines and checkout scanners to business computers and facsimile machines. It is also clear that our society has moved from a manufacturing-based work force to an information and technological rich work environment without adequately training the people who must support the environment. We also can count on technology playing an increasingly larger role in our lives.

The message is indisputable: If our students and our society are going to compete in the marketplace of the future, then our population must be scientifically and technologically literate. And there is no better place to begin to integrate the tools of technology than in the science classroom.

Technology as a Tool for Teaching and Learning

The integration of technology into the classroom provides students with new tools for exploration, discovery, and learning. Today it is possible for teachers and students to use computers and their applications, including word processors, data bases, and spreadsheets to write, manipulate, and show data and graphics. Additionally, interface probeware, laser disc players, and CD-ROM players can help to collect and organize data for both individuals and laboratory groups.

The computer can compile whole-class data that can be distributed for student analysis. Data can be easily entered, sorted, graphed, and reported. What is on the computer screen can be projected onto a large screen through the use of a liquid crystal display (LCD) making both teacher- and student-generated text and graphics available to the whole class.

Some of this technology has been developed specifically for the science classroom, including:

Probeware: A common procedure used to obtain data in the laboratory involves constant monitoring, reading of measurement devices, and recording—all good practices. However, the continual repetition of this procedure can be boring. The use of probeware removes the tedium of collecting data and allows students to focus their attention on the analysis of the data. Moreover, probeware increases accuracy of measurement, providing better data that can

help lead to better conclusions.

The ability to obtain data over very long or very short periods of time becomes possible with the use of the computer and probeware. Students are provided with immediate feedback. Data collected from probeware is very visual and can be presented in either table or graph form very quickly by the computer. This feature allows students to begin interpretation while the experiment is still fresh in their minds.

Videodiscs The computer can access movie or still visuals with a touch of a finger, the click of a mouse. CAV (continuous angular velocity) videodiscs provide both visuals and interactive programming. CLV (continuous linear velocity) discs can run movie clips and/or still visuals. A laser disc allows for access to any image on the disc in any order that is desired.

Interactive video combines a laser disc with Hypercard on a computer. This is an invaluable tool for investigation and problem solving. Videodisc material is easily adaptable to various teaching styles and curricula. A science class that is supported by easy access to quality graphics is a fertile environment for innovation, creativity, and curriculum ownership.

A Scenario for Incorporating Media and Technology in the Classroom

A small group of middle-level students work on a weather project designed to track weather patterns and fronts across the country. Using national weather information, they make predictions about what the weather will be at a given point in time in Massachusetts. The students have set up a weather station containing probeware that can measure temperature, barometric pressure, humidity, and wind direction and speed, as well as recording and graphing this information directly on their computers. Additionally, students make regular observations of cloud types. They keep on-going records of physical conditions and compare their results with those of the local weather bureau.

All members of the student group are responsible for information concerning the types of clouds and precipitation that are associated with specific fronts, and with what speed different kinds of fronts generally move through an area. In the library, the search for resources can be enhanced and speeded up by the use of the CD-ROM.

TOOL	USE	ADVANTAGES

Computer Utilities
Word process
Database
Spreadsheet
Graphics
	Classroom Management
	Grades
	Stores and adjusts worksheets/tests
	Lab & research reports
	Organization/analysis/manipulation/
	  graphing and evaluation of data
	Manipulation of numbers/text &
	  graphics
		Multiple modes of presentation
		Storage of vast amounts of informa-
		  tion
		Allows for large/small group or indi-
		  vidual work
		Can be used for any content area
		Does computations

Software Simulations Drill Games	Presents new information Provides review and drill Approximates real situations Allows for individual practice	Provides alternatives to dangerous/expensive activities

Liquid Crystal
Display Projector
	Projects computer screen for large
	  group viewing
		Allows large group access to individ-
		  ual computer screen

Probeware	Measures: temperature, pH, pres-
	  sure, time, light intensity, distance,
	  motion, sound, wind, speed, volt-
	  age, mass
		Provides fast, accurate measure-
		  ment which can be shown in both
		  table and chart form

Multimedia
Laser Disc Player
CD-ROM
Scanner
Video camera
Recorder
	Provides text and graphics which
	  can be manipulated as well as oth-
	  er visual and auditory resources
		Provides a library of high resolution
		  visuals and auditory resources
		  which can be accessed at will

Telecommunications Bulletin Board Modem	Collection of real time data Communication with primary sources	Enhances the study of current issues

Cross-references and a search are done so that all perti-nent information can be selected. Print materials are used to get an initial overview of the subject.

One student accesses information using a laser disc player that can show slides and movie clips of different kinds of clouds moving past an area with a particular weather pattern. Another student uses the encyclopedia on CD-ROM in the library to explore how clouds relate to specific fronts and weather patterns.

The students also use their electronic network to telecommunicate with middle-level students working on a similar project in Ohio. The Ohio students provide the students in Massachusetts with real-time data as a front moves through their area. The Massachusetts students track the front and make a prediction as to when this weather pattern will reach them.

The students use the word processor and data base on their computer to tabulate their data and to write back-ground information on the correlation between cloud types and fronts. They take images that they have scanned into their computer of national weather maps from local newspapers to show their classmates the movement of the front. They use an LCD projector hooked up to their com-puter to show their data and images to the class. While they are presenting information on particular cloud types, they access photos and dynamic moving footage of cloud movement from their Earth Science Laserdisc. All of this information can be accessed in any order because of the Hypercard™ format on which it has been placed.

Getting Started

It is not necessary to have all the equipment mentioned in the scenario to begin to use technology in the classroom. A single computer will get you off to a good start. Nor is it necessary for each student to use every application all the time. What is important is to expose students to these appli-cations as often as it is scien-tifically and educationally ap-propriate to do so.

Technology for Lifelong Learning and Problem Solving

Skills that can be taught in a technologically enriched sci-ence classroom, such as prob-lem solving, creative thinking, and a sense of open inquiry are becoming increasingly impor-tant. To succeed in the infor-mation age, students must be introduced to the skills they will need in their future workplaces. The skills that we ex-pose our students to today will be the foundation on which they build a career and continue to grow in the future.

The tools of technology foster independence, coopera-tion, communication, and the ability to organize, manipu-late, and evaluate data and to use multiple resources. These are skills that encourage and empower our students to become lifelong learners and problem solvers. ■

From Theory to Practice

Prentice Hall Science . . .

• Is an integrated learning system with a variety of technology components including interactive videodiscs/CD ROM, interactive videodiscs (Level III), videodiscs, and computer test banks (IBM, Apple, or MAC).

• Integrates all components in the teaching strate-gies in the *Annotated Teacher's Edition* at the point of use, where they directly relate to the sci-ence content.

• Uses the technology as an alternate way to pre-sent and involve students in learning science concepts.

• Allows teachers the opportunity to bring events into the classroom via technology and create a learning environment that would not be possible.

TEACHING TO DIVERSE CLASSROOM NEEDS

Science teachers today have the task of accommodating a much broader range of student abilities, interests, and needs than ever before. For not only does the typical science classroom include students of different ability levels, it can also include mainstreamed students who have mental or physical disabilities, such as emotional handicaps, learning disabilities, or physical impairments. Further complicating matters, recent research suggests that individual student learning styles vary greatly. The following suggestions and guidelines are designed to help teachers meet the formidable task of creating a learning environment in which all their students can thrive.

Accommodating Different Learning Styles

Educational researchers have isolated at least seven different personal learning styles—cognitive and behavioral patterns people use while

> *A variety of learning approaches and activities must be employed.*

learning. The linguistic learner, for example, learns best by saying, hearing, and seeing words; the logical/mathematical learner by categorizing, classifying and working with

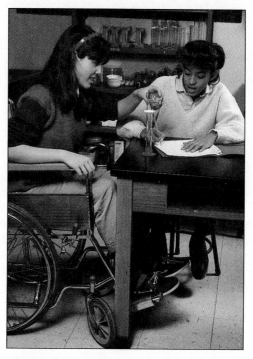

abstract patterns and relationships. Spatial learners learn by visualizing and working with colors and pictures; kinesthetic learners grasp concepts by touching and moving. Interpersonal learners master material by sharing, comparing, relating, and cooperating, while intrapersonal learners prefer to work alone at their own pace on individualized projects. Musical learners learn best when rhythm, melody, and music are involved. In addition, some of these learners move from the whole to the part, learning best when examples precede concepts. Others move from one part to the whole.

With some or all of these diverse learning styles present in a science classroom, the message to teachers is

clear: A variety of learning approaches and activities must be employed. Thus, a flexible program that combines lecture, demonstration, oral reading, discussion, and student hands-on investigation is essential. Additionally, a mixture of learning tasks or activities should be employed within each class period. Opportunities for manipulative learning activities should be provided as often as possible and textbook reading and investigations should be supplemented with audiovisual aids. In this way, all students are given an equal chance to succeed.

Teaching "Challenged" Students

In an effort to provide mentally and physically challenged students with the least restrictive learning environment possible, many such students are being mainstreamed into nonspecialized classes. By being aware of the unique needs of these students and by making a few adjustments, teachers can help these students maximize their learning experience.

Students With Learning Problems

Learning processes that include inferences and abstract reasoning are often more difficult for students who have learning problems. Such students include those who have some degree of mental retardation. In order to help these students grasp facts and concepts in science, it is important to provide daily learning goals at a pace that will allow the goals to be achieved. These students will benefit greatly from the use of concrete examples in the classroom that relate back to daily life. The need to reinforce lessons is also important for such students. Moreover, since many of these students will have experienced failure in their studies, it is vital to provide as much positive reinforcement as possible. Emphasize success and minimize failure whenever possible.

Students With Visual Problems

Students who are blind, as well as those with limited sight, are more dependent on senses such as hearing

than other students. As a result, these students should always be seated where they will be able to hear the teacher and their classmates most easily. Tape recording lessons will help these students study and go over material at their own pace. Also, classmates can be a great aid by providing descriptions of photographs and illustrations in the text.

Students With Hearing Problems

Students with hearing problems are far more dependent on the written word than other students. Usually, these students should be seated near the front of the room so that they can read the teacher's lips. The teacher should enunciate every word and avoid talking too quickly. All instructions and assignments should be written down for these students and visual aids should be used liberally. Allow students who cannot hear well to copy the notes taken in class by classmates.

Students With Other Physical Problems

Students who have physical problems that require crutches or wheelchairs will need extra room to get around in the classroom. Take care to make sure such students do not try to stretch their limits beyond their physical capabilities, but do not treat them any more differently than necessary so that they will feel an integral part of the class.

Students who have physical problems due to disorders such as muscular dystrophy, or other disorders that deter motor coordination, will often have trouble in the laboratory setting. Holding flasks, pouring liquids, and using other equipment may be beyond their capabilities. If these students can write, it is often best to assign them the task of recording during investigations while their lab partners carry out the more physical aspects of the investigation.

Some students may have illnesses such as diabetes or epilepsy. In general, these students will not need any special care. However, the teacher should be aware of any special problems or symptoms these illnesses might present in order to obtain prompt medical attention when necessary. ■

From Theory to Practice

Prentice Hall Science . . .

• Creates an ideal environment for students with all kinds of learning styles by offering a variety of multimedia and technology components.

• Helps students having reading difficulties with the English Guide for Language Learners available in print and audiotape.

• Provides a variety of learning approaches and activities.

• Offers many opportunities for hands-on learning.

• Includes a highly visual presentation in the student text as well as full color transparencies to use as teaching aids.

WHAT'S THE BIG IDEA?

Literature in a Science Textbook!

What better way to open up the world of science to students than by exposing them to it in quality literature! Why literature? Great literature represents a world of knowledge and experience. It provides students with insights into their own lives and the lives of others. It helps children discover how people think and respond to life's everyday challenges. The language is rich, fresh, and engaging. It informs, instructs, and surprises as it entertains.

Through a well-constructed story, literature can provide a context for the concepts presented in the science textbook, thereby enhancing, expanding, and enriching that content. Students will interact with scientific information, not as isolated facts or skills, but in a more holistic way. Quality literature can help children discover that science is much more than a storehouse of concepts, data, experiments, and statistics. Through literature, students can meet scientific thinkers, such as Benjamin Franklin and Dr. Daniel Hale Williams, in a personal way. They can explore scientific careers through the eyes of astronauts, conservationists, paleontologists, and authors who care about the world around them.

Activity (Reading) is an exciting new feature of *Prentice Hall Science*. What is it? How will your students benefit from it? Briefly, *Activity (Reading)* identifies outstanding trade books and invites students to investigate areas of science incorporated in these titles. We have selected the titles to support concepts in Life Science, Physical Science, Earth Science, and the Human Body. There are selections in every textbook—each selection strategically placed. Students can call upon prior knowledge as they read the selection, thus ensuring their comprehension and enjoyment. Or they can have their reading appetite whetted by an intriguing title.

Great literature can be a catalyst for developing and integrating science language skills. After reading a literature selection, students can be encouraged to interact with the story through questions to think about, talk about, and write about. Others' ideas about the literature can also be discussed. Further literature reading is encouraged with a bibliography called For Further Reading found in the Appendix.

The selections chosen for *Prentice Hall Science* have been written by award-winning authors. Many of the selections are books honored as Outstanding Science Trade Books for Children. Others have been honored for excellence with the Newbery Medal, the Caldecott Medal, the Boston Horn Book Award, and the American Library Association Notable Children's Book Award.

We at Prentice Hall strongly believe that students benefit greatly from exposure to literature. Thus we have extended an invitation to read in our *Prentice Hall Science* Program. ■

From Theory to Practice

Prentice Hall Science . . .

• Encourages the reading of literature with Activity (Reading) activities throughout the texts.

• Provides three magazine-style articles at the end of every book focusing on Adventures in Science (current scientists), Issues in Science, and Futures in Science. Each article has a corresponding reading skill worksheet in the *Activity Book*.

• Suggests additional literature selections in a bibliography called For Further Reading preceding the Appendices.

Writing in Science

Joan Develin Coley

"Writing across the curriculum" sounds like one of those catchy phrases favored by schoolboards—a notion here today, gone tomorrow. Surprise! The truth is that "writing across the curriculum" shows every sign of remaining with us permanently. That's because it's one of the most powerful educational tools available. In order to understand its power, let's think about the shift in writing instruction over the last few decades.

Writing, Past and Present

We used to steer student writers toward focusing on the product. Children learned particular modes of discourse, often by emulating models of writing that had little to do with their own interests.

Now, in dramatic contrast, we emphasize the dynamic process that writers go through. Very often, we have students choose topics on the basis of personal meaning and relevance to their own lives. And we pay a lot more attention to something called prewriting: before putting pen to paper, students gather ideas through reading, small-group and class discussions, and hands-on activities. Then, as they compose, youngsters merge new information with background knowledge. Writing, in fact, is a highly effective way for children to practice combining new data with prior knowledge.

Science Writing: Hidden Benefits

If the main goal of a science program is to get children to understand, remember, and apply science concepts, there's probably no more powerful tool than having students "write science." Research indicates that writing about science improves not only science vocabulary, but also performance in reading and thinking in the discipline.

It's clear that students understand science concepts to a greater degree when they write about them. In order to write, they have to process information they've picked up through reading discussions and hands-on activities. They reflect on ideas and reorganize them. In effect, students rehearse the material once again—sometimes many times—as they write. Reflection, reorganization, and rehearsal all contribute immensely to learning and retaining information.

Writing in science accomplishes even more! Composing prose in a content area activates critical thinking skills. Science writing, which requires the application of science concepts to novel situations, offers many opportunities to compare, classify, synthesize, and process information. These critical thinking skills automatically come into play as children restructure the ideas they've gathered into new forms built with their own words.

It's important to provide varied writing tasks in science, because each type of writing promotes different kinds of thinking and learning.

- Writing a summary encourages comprehension and retention of key concepts.
- Writing to compare or classify pieces of information sharpens children's understanding of the relationships between things.
- Writing to persuade activates still other thinking skills, such as recognizing and synthesizing information.

Naturally, writing in science also improves general writing fluency. Proficiency in any skill develops with practice. And the best way to promote lots of science writing is to generate a rich supply of prewriting input to help children come up with ideas they really want to express.

If we want students to grasp and retain science concepts, why not use every means toward this end? We know that writing is an educational power tool. Given our current understanding of the ways children assimilate and retain information, "writing across the curriculum" is a phrase we'd do well to add to our own vocabularies—and put into frequent action in the science classroom. ■

From Theory to Practice

Prentice Hall Science . . .

- Motivates students to write creatively and explore their prior knowledge with a Journal Activity at the beginning of every chapter.

- Provides further writing opportunities with Activity (Writing) activities throughout the texts.

- Includes a Using the Writing Process question in every chapter review.

- Offers further writing assignment suggestions on the chapter review pages in the *Annotated Teacher's Edition* called Issues in Science.

- Encourages the writing and sharing of activity results.

Today's science classrooms are meeting and learning places for students with a wide variety of cultural and ethnic backgrounds. This diversity presents both an opportunity and a challenge. The opportunity is the rich cultural variety that students bring with them to the classroom in terms of their customs, traditions, and experiences. The challenge is meeting the unique individual needs of each student so that all students have an equal opportunity to gain an understanding of the basic scientific principles that are so much a part of our lives.

This then is the basic goal of a multicultural orientation to science teaching. By making what goes on in the science classroom more relevant to the needs of students from diverse backgrounds, more students will be attracted to the sciences. More students will become scientifically literate and more students will be inspired to pursue careers in the sciences and technology. In essence, teachers who are aware of the multicultural approach to teaching science stand a better chance of stimulating scientific curiosity and sparking a desire to learn in all their students.

What Is a Multicultural Orientation to Science?

Multicultural awareness in the science classroom involves two fundamental realizations. The first is an understanding that the discipline of science has a culture in and of itself. Moreover, there are behaviors and attitudes that scientists consider vital to successful scientific inquiry which may be at odds with the values of some cultures. Some of these attitudes include:

Curiosity Scientists ask a lot of questions. Any observation that they make may be the catalyst for an investigation.

Open-mindedness Scientists endeavor to maintain an unbiased view of the world, to base their decisions upon acquired data, and to refrain from making judgments until sufficient data is obtained to warrant a conclusion.

BUILDING MULTICULTURALISM INTO THE SCIENCE PROGRAM

Steven J. Rakow

Skepticism Not only do scientists ask a lot of questions, but they also question a lot of answers.

Students who have been reared in a culture that values "being seen and not heard" and rejects any possibility of questioning elders may find these attitudes difficult to adopt. Thus it is necessary for the science teacher to be sensitive to such cultural traditions and how they may impact on science learning.

The second is an awareness on the part of students of the vital role that science plays in society. From the mass media to mass transit, from the exploration of outer space to the probing of the inner spaces of the human mind, science and technology have made an impact upon all aspects of our lives. Once students grasp this relevance, science concepts will spring to life before them.

How Can Teachers Build Multicultural Awareness?

There are many things teachers can do to make science more accessible to their diverse population of students. Here are some specific suggestions:

1. *Start from common, concrete experiences.* Students from varying cultural backgrounds bring diverse experiences to the classroom. This can bring a great richness to the class. It can also challenge the teacher to find a common ground of experience for the students. Providing students with common, concrete experiences allows them to begin the lesson at the same place. For example, a lesson on genetics might start with the teacher asking students to look around the classroom and notice all the ways in which each individual student's appearance is unique. It could be mentioned that the very fact that we can recognize each other is due to our genetic diversity. The teacher should also point out that although people may differ dramatically in height, weight, eye color, hair color, and skin color, as human beings we are all more alike than different.

2. *Provide real-life applications of science.* Teachers should show ways in which the science that students learn has an impact upon their lives. From the technology of video games to advances in health care to the fiber-optic networks that make it possible for students to telephone their friends, science has a direct connection to students' daily lives. Teachers who relate their science lessons to students' hobbies, interests, work, and career plans find that student motivation for learning increases dramatically.

3. *Provide examples of science from various cultures.* Science teachers should help to dispel the myth that science is just the domain of white males from North America or Northern European countries. Important scientific advances have happened all over the world, from the Chinese invention of gun powder (possibly as early as 100 A.D.) and the compass to the so-

phisticated astronomical observations made by the Mayan cultures of Central America.

4. *Cite examples of role models from different cultures.* By providing students with examples of individuals from their culture who have made contributions to science, students are better able to form an image of themselves as a potential future scientist.

One such example is African-American physician Dr. Daniel Hale Williams, who performed the first successful open-heart surgery in the United States nearly one hundred years ago. Students should be encouraged to explore the lives and achievements of these important scientists.

In addition to famous scientists, teachers can draw attention to members of the local community who hold jobs in science-related fields. Inviting local scientists and engineers into the classroom to speak about careers in science can also be a great way to make connections between business and industry in the schools.

5. *Provide examples of the impact of science upon society and our way of life.* When students begin to see that science affects them and touches upon almost everything that they do, they gain a better appreciation of the richness

of the discipline and are less likely to view science as just another requirement for graduation.

To help students grasp the extent to which science shapes our culture, have class members list the machines that might be found at home. Answers could include radios, televisions, VCRs, toasters, lawn mowers, automobiles, and so on. Then ask them to think about how their lives would be different without each of these inventions. Teachers also could ask students to imagine possible future scientific advances (for example, space travel for the general public, medical discoveries that lengthen the average life span to 120 years, and so on) and how they might change our lives in the coming decades.

Science permeates every aspect of a student's life. A multicultural orientation to science teaching merely acknowledges the diversity of the students who are part of our classes and attempts to make what they learn in the science classroom relevant to their daily lives. When science education is capable of meeting the unique individual needs of each student, then we truly achieve the goal of science literacy for all Americans. ■

From Theory to Practice

Prentice Hall Science . . .

• Integrates cultural richness—customs, traditions, and experiences—into the content and visuals.

• Uses examples of science, as well as contributions to science, from different cultures.

• Provides real-life applications of science in the content, visuals, and activities.

• Offers role models from various cultures in the content and, in particular, the Adventures in Science article in the Gazette section of the texts.

• Suggests literature selections from various cultural backgrounds in the Activity (Reading) activities and For Further Reading section.

• Begins every chapter section with a Multicultural Opportunity teaching strategy in the *Annotated Teacher's Edition.*

Creating a Positive Learning Environment for Students with Limited English Proficiency

Pat Hollis Smith

Adjusting to a new environment, its culture, and language can be an overwhelming experience for anyone, much less a young person who faces the possibility of failure in the classroom, not because of a lack of innate potential, but from inadequate English-language skills. This deficiency creates a major learning problem for limited English proficiency (LEP) students. They must begin school in whatever grade level their chronological age places them and compete in a language with which they have little or no background. Add to this dilemma in some cases a strong bond with their ancestors' zeal for outstanding achievement and you have a capsulized view of the predicament in which many LEP students find themselves as they begin their transition into life in their adopted homeland.

However, with the aid of concerned teachers who are sensitive to what these newcomers are experiencing, LEP students eventually can become acclimatized and, over time, begin to "bloom where they are planted." Once vocabulary and pronunciation are mastered, even the most complex concepts can be grasped by students who previously suffered frustration and anxiety from lack of understanding and communication skills.

LEP Students in the Science Classroom

The study of science presents an especially tough challenge for LEP students. Because scientific terms and concepts sometimes present problems even for those who are fluent in English, it is crucial for teachers to be aware of the dilemma of those with limited English proficiency and to take steps to create a learning environment in which they can flourish. The following list of recommendations is designed to help teachers create an optimal learning environment for LEP students, one which affords them the opportunity they need to be successful in today's science classroom.

Ways to Help LEP Science Students Bloom

- Seat LEP students near the front of the classroom or in an area where there are few distractions.
- Learn the correct pronunciation of students' names, taking care not to anglicize them.
- Speak slowly, enunciating clearly and maintaining a normal tone of voice. Simplify language whenever possible and rephrase sentences for clarification, if necessary.
- Provide clear illustrations and concrete examples using pictures, models, transparencies, etc. Label all props until students can readily identify them.
- Prepare cassette tapes of each chapter that students can use for independent or small-group listening study and review.
- Provide newspapers and magazines when articles are required for assignments. (As English is not usually spoken in the home of LEP students, English-language periodicals are not available.) Also, be aware that students may need library assistance with research projects.
- Use audio-visuals as often as possible to assist in student comprehension of complex concepts and skills and to reinforce vocabulary.
- Utilize remedial and average vocabulary/reading skill-building materials found in the resource material accompanying the text.
- When designating groups for cooperative learning activities, place no more than two LEP students of the same nationality in a group.

- At the teacher's discretion, oral testing may be done for new arrivals. Use positive reinforcement often to help boost students' self-esteem which often regresses during transition.
- Provide students with a copy of the following sample page. Using this page as a guideline, require them to list every unfamiliar term they encounter in each chapter (using both manuscript and cursive handwriting since their native language may not use the Roman alphabet). These lists can be kept in a notebook as students' personal glossary for the course.

CHAPTER 1 VOCABULARY		
Term	Pronunciation Native Language	Definition
tadpole/tadpole	tad'pōl con nòng-nọc	a very young frog

- Where available, enlist the aid of your school's ESOL teacher(s) in helping students complete and study their lists. If this is not possible, model the correct pronunciation and provide both a written phonetic version and a simplified definition for each term. When appropriate, include a simple illustration.
- Allow students to use their native-language dictionary to find the translation of a term or, after the above help is given, to write the word in their first language from memory. ∎

From Theory to Practice

Prentice Hall Science . . .

- Offers a *Spanish Guide for Language Learners* and an *English Guide for Language Learners* available in print and audiotapes.

- Supports vocabulary and concepts with dynamic visuals in the student texts.

- Begins every chapter section with an ESL Strategy in the *Annotated Teacher's Edition*.

- Provides full color transparencies of key ideas from the text to help students visualize the concepts.

CONCEPT MAPPING:
Making Learning Meaningful

Karen K. Lind

Concept maps make key ideas abundantly clear. They are, in essence, visual road maps showing pathways that meaningfully connect major concepts. In a concept map the hierarchial nature of concepts is visually emphasized, with the superordinate-subordinate relationships between key concepts distinctly shown. Thus, students who have difficulty grasping such relationships when they are presented in a written or oral format are often able to perceive these connections upon viewing a concept map, which has dramatic visual impact.

Humans naturally think in a hierarchial way. We classify objects and ideas into categories to assimilate them in our minds. Concept maps, by their very nature, utilize this natural inclination to relate and develop meaningful concepts. Because of this, concept maps can enhance teaching and learning. For example, concept maps can be used effectively to demonstrate the relationships between concepts in a lesson or unit. One method that works well is for the teacher to construct a poster or wall-sized concept map. This can be made using large rectangular pieces of oaktag, colored felt pens, and

yarn. The map is kept in full view of the class as work on the lesson or unit progresses, and is modified as meaning develops. This is only one of the many ways concept maps can be used to clarify and assist instruction. The following suggestions and ideas provide a wide range of applications in the science classroom.

Using Concept Maps in Preinstruction

Constructing a class concept map before the lesson begins can make the task of determining what a learner already knows easier and can provide a conceptual benchmark from which students can construct meanings. As the class sketches out the key ideas before reading a section, the teacher gains valuable insight into each learner's cognitive structure in regard to a specific topic.

Additionally, as students build a concept map relating to the concepts in the textbook, they can incorporate their own past experiences and prior knowledge into the classroom learning. In this way not only is material made more meaningful for students, but the process of concept mapping enables teachers to pinpoint and clarify student misconceptions.

Sharing, Thinking, and Negotiating Meaning

A team effort in concept-map making affords students an opportunity to exchange views and recognize missing links. Because of the individual nature of concept formation, students relate meanings in different ways. Working in groups of two or three to construct a concept map provides students with a format for sharing their ideas, negotiating the meaning of specific textbook passages, and adding to the meaning presented in the textbook, if necessary. Experiences that the student brings to the lesson are emphasized and integrated into the concepts being developed. Students gain experience in cooperating and sharing, and teachers have an opportunity to guide and participate in the learning process.

The Year at a Glance

To help students relate to the concepts that they will be studying over the course of a year or semester, teachers can make and post a large global concept map of the proposed instruction. To illustrate meanings and facilitate recall, teachers should encourage students to add photos, magazine clippings, and drawings of the key concepts. Students can also make detailed maps as they progress through the year. In this way, students will be able to perceive connections between key concepts and the year's study as well as to orient themselves quickly and know where they are headed.

Reading Newspapers and Magazine Articles

A concept map can create a framework for understanding the meaning of the ideas presented in a newspaper or magazine article. It can also make the substance of the reading easier to recall. After students have read an article, have them read the article again and circle the main concepts. Then ask them to construct a concept map representing the key ideas in a hier-

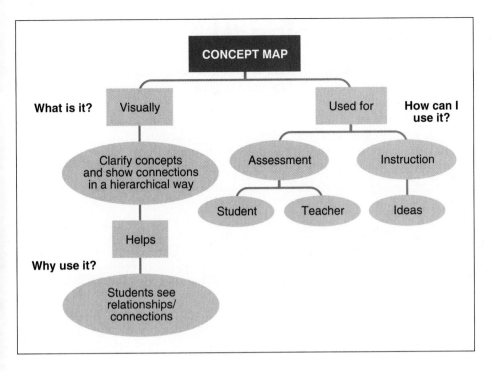

levels of thinking presented in Bloom's Taxonomy—knowledge, classification, and comprehension—are relatively easy to construct. However, questions and problems that indicate whether students are operating on the higher levels described in Bloom's Taxonomy—analyzing, synthesizing, evaluation—are more difficult and time consuming to design. A major benefit of having students make concept maps is that the activity of constructing a map requires students to operate on all six levels of Bloom's Taxonomy. Moreover, these levels can be assessed before learning (as a diagnostic assessment), during learning (to modify instruction), and after learning (as a schematic summary of what has been learned).

Concept maps also provide students with a means of self-assessment. Students can compare the concept maps that they or the class have constructed at the beginning of a lesson with the ones that they make at the completion of the lesson. The students can see for themselves what they have achieved.

When concept maps are used in the science classroom, students have the opportunity to identify key concepts and relate them to prior knowledge. Teachers can readily organize concepts for student discussion and clarify misconceptions that may become apparent. Both teachers and students are able to assess learning and participate in curriculum planning. By working together to relate concepts, learning becomes more meaningful for both teachers and students. ∎

archial order. If the article is short, such as a summary of an expert's work, concepts and phrases needed to develop meaning may have been omitted. Have students add any key ideas they feel have been left out of the article.

Writing or Presenting Ideas

Even professional writers can be intimidated by a blank sheet of paper. Concept mapping is one way students can ease into a writing project. Have students begin by listing a few concepts they want to include in the paper. They can then construct a brief map to guide the first few paragraphs. While a complete concept map probably cannot be developed prior to writing a paper, the initial concept map can be modified as the paper progresses. Most importantly, concept mapping will get students off to a comfortable start. The creation of posters, notices, exhibits, and so forth, can also be approached in this manner.

Curriculum Development

Concept maps can be used to help teachers identify the most important concepts to be covered during the year. With a clear road map, a quick

glance is all that is needed to keep on track. Sharing the plan for the year is also a good way to include students in the planning process. Students' prior experiences and misconceptions may lead to an emphasis of one concept over another. A concept map will keep the most important concepts in focus. Student and teacher assessment of progress may also lead to major modifications of instruction to reinforce the main concepts.

Concept Maps in Assessment

The use of concept mapping as an assessment tool provides teachers with additional strategies for assessing student understanding of scientific concepts and of the way concepts relate to each other. Student-designed concept maps enable teachers to assess higher-level objectives, such as those described in Bloom's *Taxonomy of Educational Objectives*. For the most part, examination items and questions that assess the lower

Concept Mapping #1

Concept mapping is similar to taking notes, and it will help you to keep track of the main ideas and supporting details as you read. To make a map of your reading, find the main idea. Write the main idea in the blank circle below. As you identify other ideas or information that supports the main idea, add these to lines connected to the main circle. You can add as many lines as needed to complete your map. Your map should look something like the example in the upper right hand corner. When you are done, your map will contain the most important points in your reading.

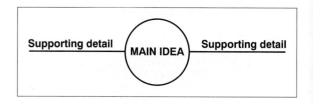

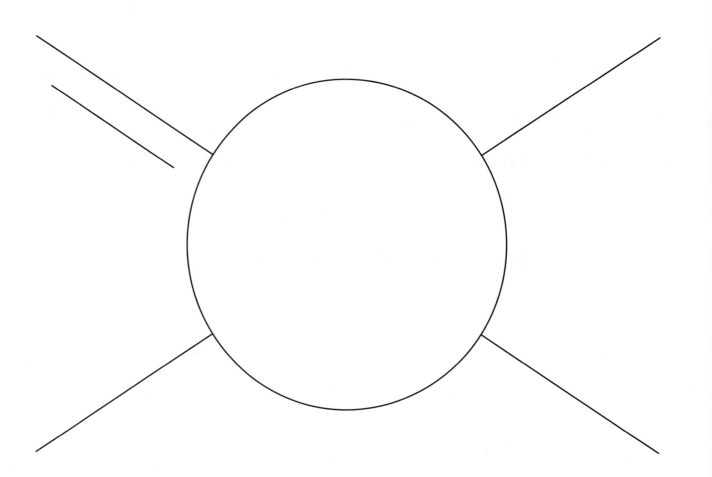

Concept Mapping #2

This circle map enables you to visualize how things change as they pass through natural cycles. As you read, identify separate stages of cycles as they are described. In the center circle, write the name for the cycle you are mapping. Then label the circles 1, 2, and so forth. Next write the name of the first stage of the cycle (see example) and a few key words describing what happens during that stage. Continue labeling the circle for all the stages in the cycle you are mapping.

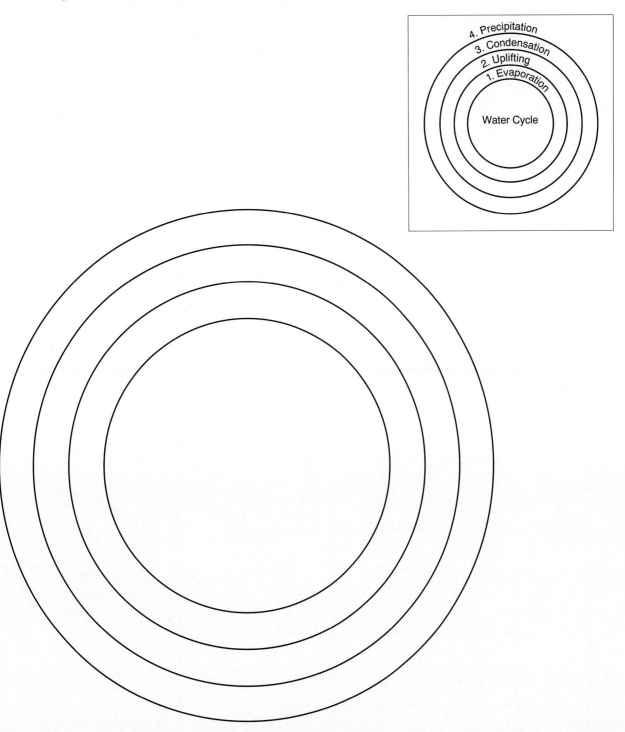

Concept Mapping #3

This type of concept map helps you to visualize and to explain cause and effect relationships. As you read, identify any causes and effects and write them in the diagram (see example). This type of map will help you to make connections and see relationships, as well as to clarify complex concepts.

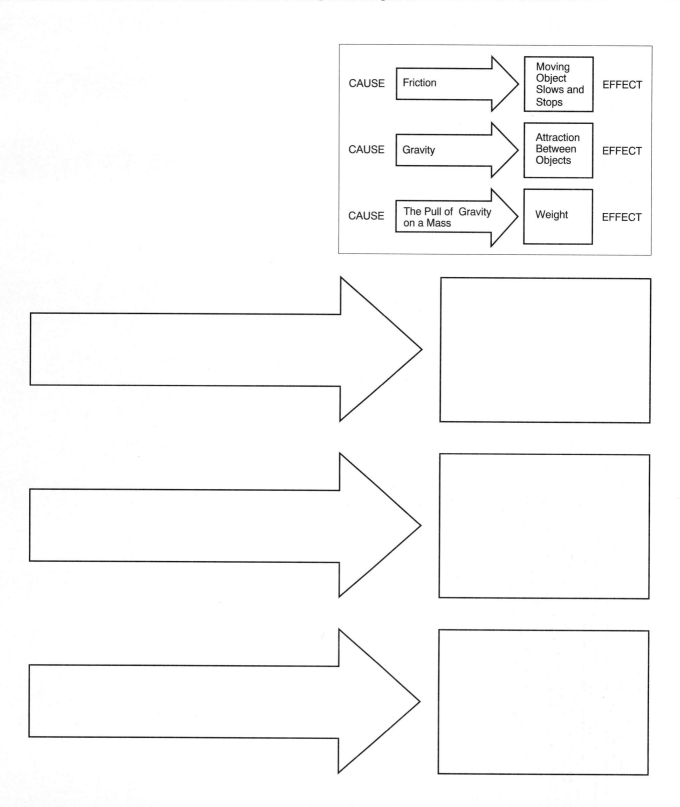

Concept Mapping #4

This concept map organizes examples of key terms as you read or review. Write the key term, question, or objective for reading in the main box (see example). As you read or study, search for examples that describe or explain the term, question, or objective in the main box and write them in the smaller boxes.

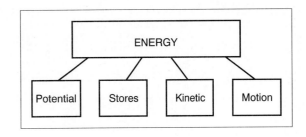

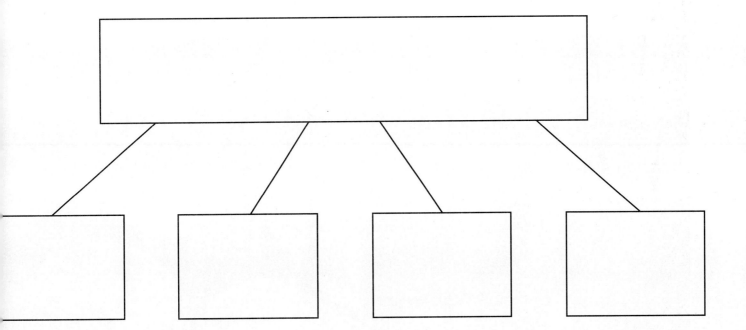

CLASSROOM RESOURCES

Any successful professional must be resourceful. This is even more
true for teachers, who must have many resources available to deal with the diversity
of students and situations in their classrooms. To help meet this need some helpful
resources appear on the following pages. They include Science Safety Rules, a Field
Trip Guide, Recipes for Common Laboratory Solutions, summaries of Project 2061
Science for All Americans and the National Science Teacher's Association Scope,
Sequence, and Coordination Project, and Graphing Skills blackline masters.

SAFETY IN THE SCIENCE CLASSROOM

Hands-on investigations, experiments, and field trips are an important—and fun—part of science learning for students. They give youngsters the opportunity to observe and actively explore scientific principles and phenomena. Your role during these activities is not only to stimulate the scientific curiosity of your students and help them learn first-hand about science, but to safeguard student safety by making sure class members follow correct safety procedures. The following suggestions are designed to help you ensure safety as your students investigate science both in and out of the classroom.

Safety Guidelines in the Classroom and Lab

Do's

- Become familiar with all federal, state, and local regulations pertaining to school safety and science activities.
- Regularly check the classroom environment to make sure that equipment and materials are in good working order and stored properly and that all hazardous materials are locked away. Also make sure that work areas are well ventilated.
- Learn the specific safety precautions for the equipment and materials being used in any given activity and discuss them with the class before the experiment begins. Also secure students' long hair and any loose-fitting clothing with ties or pins.
- Before beginning any experiment, ask students to make a list of what could go wrong that could cause injury. Then ask them to make a list of what to do should the accident occur.
- Require students to wear protective goggles whenever they work with chemicals, burners, or any substance that might get into the eyes as many lab materials can cause eye injury or even blindness.

- Advise students that any accident or injury should be reported to you immediately, no matter how minor it may seem. Know first aid measures for treating burns, cuts, and bruises and for such specific accidents as spilling acid on the skin (rinse the affected area with lots of water).
- Make sure a fire extinguisher is readily available in the event of a fire and know how to use it. Make sure students know evacuation procedures.
- Inform students of the proper way to use and handle glassware and sharp instruments.
- Tell students to carefully read the directions for an experiment two or three times before beginning and to follow the directions exactly as they are written.
- Immediately wipe up any liquid spills on the floor to prevent falls.

Don'ts

- Never permit students to undertake any science investigation without your supervision.
- Caution students never to mix chemicals "for the fun of it." It might produce an explosive reaction that could cause serious injury to themselves and others.
- Warn students not to taste, touch, or smell any chemical that they don't know for a fact is harmless.
- Tell students never to heat any chemical unless they are specifically instructed to do so since some chemicals that are harmless when cool become dangerous when heated.
- Caution students never to heat a liquid in a closed container as the expanding gases may blow the container apart.
- Tell students not to tilt a test tube toward themselves or anyone else or even hold it upright, but always to tilt tubes away from themselves and others.

- Advise class members not to touch any part of the the face or body when performing an experiment with chemicals, plants, or animals and to wash their hands thoroughly when the activity is completed.
- Warn students against ever performing an experiment for which they do not have written instructions either in a text or from you.

Tips for a Safe and Successful Field Trip

The success of a field trip depends on careful preparation, both by you and your students. The following recommendations are presented to help you organize and conduct a safe and productive field trip.

1. **Site Selection** It is up to you to select an appropriate field trip site for your students, depending upon your locale. To help you make the necessary arrangements, we suggest you keep a school file of appropriate sites in your area. The file should contain information about who to contact at a site, directions to or a map of the site, fees (if any), the hours the site is open to the public, and availability of meal facilities and restrooms.

Once you have selected a site, it is wise to visit it prior to the field trip to inspect the facilities. While there, you should locate and inventory work-study areas, make a list of specific equipment your students will need, and note the site restrictions and danger spots. If any of your students are physically challenged, take stock of the site's facilities for the handicapped.

2. **Planning the Trip** Request written permission for the trip from school personnel and keep the written permission on file. Notify each student's parent or guardian in writing of your destination, departure and arrival times, mode of transportation, and necessary expenditures. Correlate the trip with your lessons and text material. Tell your students the why, where, and when of the field trip and provide class time for advance research on the site. Be sure to inform students of any special clothing or equipment they will need—special shoes, shorts, hand lenses, notebooks, etc.

3. **On the Field Trip** Make a head count of your students at each boarding and departure, and periodically during the trip. Each adult should have a list of the students he or she is to supervise and should remain with that group throughout the entire trip. While traveling to the site, discuss the investigation with your students. Once at your destination, keep the group together.

Make sure students understand the purpose of the trip. Have them make sketches, drawings, plans, or maps, or take notes. Do not allow students to remove anything from its natural setting (an exception would be carefully selected items that are taken for observation and returned to their natural habitat).

Most importantly, be enthusiastic, but don't rush. Don't try to crowd too much into one field trip. Keep in contact with the individuals in the group and be alert and prepared for the "teachable moment."

4. **Follow-up Activities** The learning value of a field trip depends largely on the follow-up activities you promote. While you are returning to the school, have students exchange ideas and discuss their experiences and observations at the site. Encourage students to ask questions and propose future activities related to the field trip. Schedule individual or group reports and have students evaluate the trip. You may also want to have students prepare exhibits or displays using their sketches, maps, or other materials from the trip, or have them use the library to investigate questions arising from the trip. ■

FIELD TRIP GUIDE

Representative places of interest in each state and the District of Columbia are listed below. For information about other places of interest check *The Official Museum Directory* (New York: Macmillan), which is updated annually and available in most public libraries. This directory provides the address, types of collections, and facilities for each institution.

Alabama

Anniston Museum of Natural History
4301 McClellan Blvd.
P.O. Box 1587
Anniston, AL 36202
Habitats, fossils, rocks; Alabama cave; programs for children

Birmingham Botanical Gardens
2612 Lane Park Rd.
Birmingham, AL 35223
Orchids, rhododendrons, wildflowers; Touch and See; no charge

University of Alabama
State Museum of Natural History
P.O. Box 5897
Tuscaloosa, AL 35487
Fossils, minerals, plants; programs for children

Alaska

Pratt Museum
3779 Bartlett St.
Homer, AK 99603
Animals, plants; marine aquarium, botanical garden; programs for children, no charge

University of Alaska Museum
907 Yukon Drive
Fairbanks, AK 99775
Birds, fossils, mammals, plants, guided tours for schools

Arizona

Arizona Mineral Museum
Mineral Building
State Fairgrounds
Phoenix, AZ 85007
Gems, minerals, rocks no charge

Arizona Museum of Science and Technology
80 North 2nd St.
Phoenix, AZ 85004
Health, energy, physics; interactive; exhibits

Desert Botanical Garden
1201 North Galvin Pkwy.
Phoenix, AZ 85008
Cactuses and leaf succulents; arid land plants of the world

Arkansas

Little Rock Zoological Gardens
1 Jonesboro Dr.
Little Rock, AR 72205
Birds, fish, mammals, reptiles; guided tours

The University Museum
University of Arkansas
Museum Building
Fayetteville, AR 72701
Fish, reptiles, natural history

California

Cabrillo Marine Museum
3720 Stephen White Dr.
San Pedro, CA 90731
Marine plants and animals, shore birds, fish, fossils, shells, whales; no charge

The Exploratorium Museum of Science and Art
3601 Lyon St.
San Francisco, CA 94123
Participatory exhibits that explore the physical world and human sensory mechanisms

Natural History Museum of Los Angeles County
900 Exposition Blvd.
Los Angeles, CA 90007
Animals, fossils, minerals, plants; school loan service

San Diego Space and Science Foundation
1875 El Prado
Balboa Park
San Diego, CA 92101
Astronomy and space science; interactive and audience participation exhibits

Santa Barbara Museum of Natural History
2559 Puesta del Sol Rd.
Santa Barbara, CA 93105
Exhibits focus on Pacific coast and Channel Islands; dioramas of prehistoric Indian life

Colorado

Denver Museum of Natural History
City Park
Denver, Co 80205
Birds, fossils, mammals, minerals, reptiles, rocks

Denver Zoo
East 23rd and Steele Sts.
Denver, CO 80205
Guided tours: Predator or Prey, Animal Adaptations, Animals of Africa, Endangered Species, Birds, Primates, Zoo Careers, Animals of Colorado, Animal Classification, Mammals

University of Colorado Museum
Broadway between 15th and 16th Sts.
Boulder, CO 80309
Birds, butterflies, fossils, mammals, minerals, mollusks; invertebrate fossils; no charge

Connecticut

Bruce Museum
Museum Drive
Greenwich, CT 06830
Natural science collection, marine center

The Maritime Center at Norwalk
South Norwalk, CT 06850
Aquarium, maritime museum; educational programs; special exhibits

Museum of Art, Science, & Industry
4450 Park Avenue
Bridgeport, CT 06604
Planetarium and museum; hands-on physical science exhibit

Peabody Museum of Natural History
Yale University
170 Whitney Ave.
P.O. Box 6666
New Haven, CT 06511-1861
Animals, plants, fossils; no charge

Science Museum of Connecticut
950 Trout Brook Dr.
West Hartford, CT 06119
Animals, fossils, technology; physical sciences discovery room; planetarium shows

Delaware

Delaware Museum of Natural History
4840 Kennett Pike
Wilmington, DE 19807
Birds, bird eggs, mollusks, mammals; school loan service

District of Columbia

Natural Air and Space Museum
Independence Ave. and 6th St., N.W.
Washington, DC 20560
Air and space craft; aerospace; planetarium no charge

National Museum of Natural History
10th St. and Constitution Ave., N.W.
Washington, DC 20560
Animals, fossils, plants, minerals, meteorites, rocks; no charge

Florida

Museum of Arts and Sciences
1040 Museum Blvd.
Daytona Beach, FL 32014
Regional natural history and fossils; planetarium

Museum of Science and History
1025 Gulf Life Dr.
Jacksonville, FL 32207
Natural science, physical science, planetarium

Museum of Science and Space
Transit Planetarium
3280 South Miami Avenue
Miami, FL 33133
Hands-on science and computer exhibits; planetarium and laser shows

Orlando Science Center
Loch Haven Park
810 East Rollins Ave.
Orlando, FL 32803
Life and physical sciences; planetarium shows

Science Museum and Planetarium and
South Florida Aquarium
4801 Dreher Trail North
West Palm Beach, FL 33405
Discovery hall; hands-on exhibits; outreach programs

Georgia

Fernbank Science Center
156 Heaton Park Dr., NE
Atlanta, GA 30307
Birds, insects, mammals; minerals, rocks; planetarium shows

Savannah Science Museum
4405 Paulsen St.
Savannah, GA 31405
Live reptiles and amphibians; nature preserve; planetarium

Sci Trek, The Science and
Technology Museum of Atlanta
395 Piedmont Ave., NE
Atlanta, GA 30308
Physical science; participatory exhibits

Zoo Atlanta
Grant Park
800 Cherokee Ave.
Atlanta, GA 30315
African rainforest and African Savannah exhibits

Hawaii

Foster Botanical Garden
180 N. Vineyard Ave.
Honolulu, HI 96817
14 acres tropical garden; 4000 species of flowers and plants

Harold L. Lyon Arboretum
3860 Manoa Road
Honolulu, HI 96822
Living and preserved plants; programs for children

Waikiki Aquarium
2777 Kalakaua Ave.
Honolulu, HI 96815
300 species of Hawaiian and South Pacific fish, plants and mammals

Idaho

Craters of the Moon, National Monument
Box 29
Arco, ID 83213
Plants; rocks; guided tours

Idaho Museum of Natural History
Idaho State University
Campus Box 8096
Pocatello, ID 83209
Animals, fossils, plants; minerals, rocks

Illinois

Adler Planetarium
1300 S. Lake Shore Dr.
Chicago, IL 60605
Astronomy, early scientific instruments, engineering

Fermi National Accelerator Laboratory
Kirk Road
P.O. Box 500
Batavia, IL 60510
Guided tours through part of the particle accelerator; slide show

Field Museum of Natural History
Roosevelt Rd. at Lake Shore Dr.
Chicago, IL 60605
*Animals, fossils, minerals, plants, rocks;
natural history*

Museum of the Chicago Academy of
Sciences
2001 N. Clark St.
Chicago, IL 60614
Animals, fossils, plants, rocks

Museum of Science and Industry
57th St. & Lake Shore Dr.
Chicago, IL 60637
Aerospace, chemistry, physics, technology

Indiana

Evansville Museum of Arts and Science
411 South East Riverside Dr.
Evansville, IN 47713
*Natural history, minerals, rocks;
planetarium*

Indianapolis Children's Museum
3000 M. Meridian
Indianapolis, IN 46208
*Computer lab, interactive displays, natural
history, special events*

Indianapolis Zoo
1200 W. Washington St.
Indianapolis, IN 46222
*Birds, mammals, reptiles in habitats; dol-
phins and fish in aquarium*

Iowa

Grout Museum of History and Science
West Park Ave. & South St.
Waterloo, IA 50701
*Astronomy; fossils, rocks; planetarium
shows*

Putnam Museum
1717 W. 12th St.
Davenport, IA 52804
fossils, plants; natural history

University of Iowa
Museum of Natural History
10 Macbride Hall
Iowa City, IA 52240
*Animals, ecology; fossils, plants, minerals,
rocks*

Kansas

Museum of Natural History
University of Kansas
Lawrence, KS 66145
*Birds, fishes, fossils, habitats, insects, rep-
tiles, endangered species*

Snow Entomological Museum
Snow Hall
University of Kansas
Lawrence, KS 66045
*2.85 million insect specimens; educational
programs*

Systematics Museums
University of Kansas
Dyche Hall
Lawrence, KS 66045
Animals, ecology; fossils, plants

Kentucky

The Living Arts and Science Center
362 Walnut St.
Lexington, KY 40508
Natural history; program for children

Louisville Zoological Garden
1100 Trevillian Way
Louisville, KY 40213
*Birds, mammals, reptiles in environmental
exhibits*

Museum of History and Science
727 W. Main St.
Louisville, KY 40202
Natural history; space technology

Louisiana

Louisiana Arts and Science Center
100 S. River Rd.
Baton Rouge, LA 70801
Astronomy, planetarium

LSU Museum of Geoscience
Geology Bldg., Room 313C
Tower Dr.
Baton Rouge, LA 70803
Fossils, minerals, rocks

Maine

Maine State Museum
State House Station 83
Augusta, ME 04333
Minerals, rocks, natural history; technology

The Natural History Museum
College of the Atlantic
Route 3
Bar Harbour, ME 04609
Birds, mammals, plants

Maryland

Maryland Academy of Sciences
601 Light St.
Baltimore, MD 21230
*Energy, rocks, minerals; live animals and
aquarium*

National Aquarium in Baltimore
501 E. Pratt St., Pier 3
Baltimore, MD 21202
Aquarium; marine life

Massachusetts

The Discovery Museums
177 Main St.
Acton, MA 01720
Dinosaurs, fossils, physical science

Harvard University Museums
of Natural History
26 Oxford St.
Cambridge, MA 02138-2902
Animals, fossils, minerals, plants, rocks

Museum of Science
Science Park
Boston, MA 02144-1099
*Animals, human biology, minerals, plants;
planetarium*

New England Aquarium
Central Wharf
Boston, MA 02110
2000 specimens

New England Science Center
222 Harrington Way
Worcester, MA 01604
Physical and natural science, live animals and plants; planetarium

Michigan

Detroit Science Center
5020 John R. Street
Bloomfield Hills, MI 48202
All science areas; participatory exhibits

Kingman Museum of Natural History
W. Michigan Ave. at 20th St.
Battle Creek, MI 49017
Fossils, minerals, natural history; planetarium

Minnesota

Bell Museum of Natural History
10 Church St., SE
Minneapolis, MN 55455
Animals, fossils, habitats, seeds

Science Museum of Minnesota
30 East 10th St.
St. Paul, MN 55101
Astronomy, biology, fossils, minerals, rocks

Mississippi

Jackson Zoological Park
2918 W. Capitol St.
Jackson, MS 39209
Endangered species

Mississippi Museum of Natural Science
111 N. Jefferson St.
Jackson, MS 39202
Animals, fossils, habitats, plants, aquariums

Russell C. Davis Planetarium
201 E. Pascagovla St.
Jackson, MS 39201
Planetarium shows

Missouri

Kansas City Museum
3218 Gladstone Blvd.
Kansas City, MO 64123
Fossils, rocks, technology, planetarium

Missouri Botanical Garden
4344 Shaw Avenue
P.O. Box 299
St. Louis, MO 63166
Climatron, desert plants, Mediterranean plants, educational programs

St. Louis Science Center
5050 Oakland
St. Louis, MO 63110
Natural history, geology; zoology, technology, planetarium

St. Louis Zoological Park
Forest Park
St. Louis, MO 63110
Amphibians, birds, fish, invertebrates, mammals, reptiles

Montana

Earth Science Museum
106 Main St.
Loma, MT 59460
Minerals, geology, school tours

Mineral Museum
Montana Tech
West Park Street
Butte, MT 59701
Fossils, minerals

Nebraska

Henry Doorly Zoo
River View Park
10th St. & Deer Park
Omaha, NE 68107
Live animals; zoo museum

University of Nebraska
State Museum
212 Morrill Hall
14th & U Streets
Lincoln, NE 68588
Animals, fossils, rocks; health science; planetarium

Nevada

Fleishmann Planetarium
University of Nevada
Reno, NV 89571
Astronomy; planetarium shows

Las Vegas Museum of Natural History
900 Las Vegas Blvd. N.
Las Vegas, NV 89101
Dinosaur skeletons, dioramas, taxidermy demonstrations

Museum of Natural History
University of Nevada
Las Vegas, NV 89154
Fish, fossils, live animals, reptiles, rocks

New Hampshire

Audubon Society of New Hampshire
3 Sik Farm Road
Concord, NH 03301
Live animals and stuffed birds, nature center

Science Center of New Hampshire
Box 173
Holderness, NH 03245
Ecology, nature trails

New Jersey

James A. McFaul
Environmental Center
Crescent Ave.
Wyckoff, NJ 07840
Live animals, nature walks, ecology, botany

The Morris Museum
6 Normandy Hts. Rd.
Morristown, NJ 07960
Animals, fossils, natural history

The Newark Museum
49 Washington St.
Newark, NJ 07101
Natural, physical, and earth sciences; planetarium shows

Princeton University Museum
of Natural History
Guyot Hall
Princeton, NJ 08544
Fossils, minerals, rocks

New Mexico

Institute of Meteorites
Meteorite Museum
University of New Mexico
Albuquerque, NM 87131
Meteorites, rocks from terrestrial impact craters

International Space Hall of Fame
State Hwy. 2001
Alamogordo, NM 88310
Models of satellites, items from space missions

New Mexico Museum of Natural History
1801 Mountain Rd. NW
Albuquerque, NM 87104-7010
Animals, fossils, plants, rocks

New York

American Museum of Natural History
Central Park West at 79th St.
New York, NY 10024
Dinosaurs and other fossils; human biology; ecology; space science

Discovery Center of Science
& Technology
321 South Clinton St.
New York, NY 13202
Concept-oriented science and technological exhibits; planetarium

Fire Island National Seashore
For information call (516) 289-4810
Patchogue, NY 11772
Nature Walks; ecology of Fire Island

New York Aquarium
Boardwalk & W. 8th Street
Brooklyn, NY 11224
Marine life, beach programs, hands-on programs, artifacts to handle

New York Hall of Science
Corona Park
Corona, NY 11368
Hands-on science museum, light, color, the atom

New York State Museum
Cultural Education Center
Empire State Plaza
Albany, NY 12230
Animals, minerals, rocks, technology

New York Zoological Park
Bronx River Pkwy at Fordham Rd.
Bronx, NY 10460
Amphibians, birds, mammals, reptiles, habitats

Sci-Tech Center of Northern New York
New York State Office Bldg.
317 Washington St.
Watertown, NY 13601
Biology, computers, physics; hands-on science and technology; discovery museum

North Carolina

Morehead Planetarium
University of North Carolina
Chapel Hill, NC 27514
Astronomy, meteorites

North Carolina Museum of Life and Science
433 Murray Ave.
P.O. Box 15190
Durham, NC 27704
Aerospace, fossils, habitats, reptiles, rocks

North Carolina State Museum
102 N. Salisbury St.
P.O. Box 27647
Raleigh, NC 27611
Animals, rocks, fossils, minerals

North Dakota

University of North Dakota
Zoology Museum
Department of Biology
Grand Forks, ND 58202
Birds, fish, insects, mammals, reptiles and amphibians

Ohio

Cincinnati Museum of Natural History and Planetarium
1720 Gilbert Ave.
Cincinnati, OH 45202
Animals, fossils, minerals, rocks; planetarium shows

Cleveland Museum of Natural History
Wade Oval, University Cir.
Cleveland, OH 44106
Birds, fossils, insects, mammals, minerals, plants, reptiles; botanical garden

Toledo Museum of Natural Sciences and Zoological Gardens
2700 Broadway
Toledo, OH 43609
Birds, fish, fossils, reptiles, minerals, rocks; aquarium, botanical garden

Oklahoma

Kirkpatrick Center
Museum Complex
2100 Northeast 52nd St.
Oklahoma City, OK 73111
Airspace museum, planetarium

Stovall Museum of Science & History
University of Oklahoma
1335 Asp Ave.
Norman, OK 73019
Animals, fossils, plants, minerals, rocks

Oregon

Oregon Museum of Science & Industry
4015 Southwest Canyon Rd.
Portland, OR 97221
Computer, earth, life, physical, space sciences; interacting exhibits

Williamette Science & Tech. Center
2300 Centennial Blvd.
Eugene, OR 97401
Biology, computer science, physics

Pennsylvania

Academy of Natural Sciences of Philadelphia
19th St. & Benjamin Franklin Pkwy.
Philadelphia, PA 19103
Mammals, birds, fish, reptiles, fossils, minerals

Carnegie Museum of Natural History
4400 Forbes Ave.
Pittsburgh, PA 15213
Animals, minerals, rocks, plants

Science Museum and Planetarium
20th St. & Benjamin Franklin Pkwy.
Philadelphia, PA 19103
Mathematics, space science, technology

Pittsburgh Aviary
Allegheny Commons
West Park
Pittsburgh, PA 15212
Birds, fish, plants, turtles

Rhode Island

Roger Williams Park Museum
Roger Williams Park
Providence, RI 02905
Birds, diatoms, eggs, insects, invertebrates, mammals, rocks and minerals

South Carolina

South Carolina State Museum
301 Gervais St.
Columbia, SC 29202
Natural history, science technology; discover room

World of Energy at Keowee-Taxaway
P.O. Box 1687
Clemson, SC 29633
Energy and energy technologies

South Dakota

Museum of Geology
South Dakota School of Mines and Technology
501 East St. Joseph
Rapid City, SD 57701
Animals, fossils, minerals, plants, rocks

Tennessee

Cumberland Science Museum
800 Ridley Rd.
Nashville, TN 37203
Natural history, geology, general science, planetarium

Memphis Zoo and Aquarium
2000 Galoway St.
Memphis, TN 38112
Live animals; aquarium

Texas

Dallas Zoo
621 East Clarendon Dr.
Dallas, TX 75203
Amphibians, birds, mammals, reptiles

Fort Worth Museum of Science and History
1501 Montgomery St.
Fort Worth, TX 76107
Animals, fossils, meteorites, plants, planetarium

Fort Worth Zoological Park
2727 Zoological Park Drive
Fort Worth, TX 76107
Zoo, aquarium, simulated rain forest

Houston Museum of Natural Science
1 Hermann Circle Dr.
Hermann Park
Houston, TX 71030
Animals, astronomy, fossils, minerals, space science

McDonald Observatory
P.O. Box 1337
Fort Davis, TX 79734
Astronomy, daily guided tours

San Antonio Botanical Garden
555 Funston Place
San Antonio, TX 78209
Woody and herbaceous plants

Southwest Museum of Science and
Technology, The Science Place
Fair Park
P.O. Box 11158
Dallas, TX 75223
Aerospace, astronomy, energy, human body, minerals, rocks

Utah

Hansen Planetarium
15 South State St.
Salt Lake City, UT 84111
Meteorites, moon rock space technology; planetarium shows

State Arboretum of Utah
Arboretum Center
University of Utah
Salt Lake City, UT 84112
Cactuses, dwarf conifer, oak grove, trees

Utah Museum of Natural History
University of Utah
Salt Lake City, UT 84112
Dinosaurs and other fossils, herbarium, minerals, rocks

Vermont

Discover Museum
51 Park Street
Essex Junction, VT 05452
Physical and natural science, live animals

Fairbanks Museum and Planetarium
Main and Prospect Sts.
St. Johnsbury, VT 05819
Astronomy, birds, mammals, physical science, weather science; planetarium

Virginia

Science Museum of Virginia
2500 W. Broad St.
Richmond, VA 23220
Astronomy, crystals, computers, minerals, physical sciences, rocks

Science Museum of Western Virginia
1 Market Sq.
Roanoke, VA 24011
Energy, health, minerals, natural history, physical sciences

Virginia Zoological Park
3500 Granby St.
Norfolk, VA 23504
Animals, plants

Washington

Pacific Science Center
200 2nd Ave., N.
Seattle, WA 98109
Astronomy, biology, health science, minerals, rocks, space science

Seattle Aquarium
Pier 59 Waterfront Park
Seattle, WA 98101
Marine mammals, regional and tropical habitats

Washington Park Arboretum
2300 Arboretum Dr., E.
University of Washington XD-10
Seattle, WA 98195
Trees, shrubs, plants of the world

West Virginia

Sunrise Museum
746 Myrtle Road
Charleston, WV 25314
Natural sciences, discovery center; planetarium

Wisconsin

Discovery World Museum of Science,
Economics, and Technology
818 W. Wisconsin Ave.
Milwaukee, WI 53233
Health and physical science

University of Wisconsin Arboretum
1207 Seminole Hwy.
Madison, WI 53711
Plants and animals in ecological communities

Wyoming

The Geological Museum
University of Wyoming
Department of Geology & Geophysics
University of Wyoming
Laramie, WY 82071-3006
Dinosaurs and other fossils; minerals, rocks

Wyoming State Museum
Barrett Bldg.
24th and Central Ave..
Cheyenne, WY 82002
Fossils, minerals

✓ Common Solutions

✓ Fun Recipes

✓ Substitutions

Teachers may wish to make their own materials rather than purchase them. This section tells you how to make certain materials. The quantities listed below are recommended for a class of 30 students working in 6 groups.

Common Solutions

BAKING SODA SOLUTION

Place 1 L (1 qt) of water in a plastic jar, stir in baking soda one spoonful at a time until no more will dissolve. Pour off the clear liquid. Store in dropper bottles.

BEET JUICE INDICATOR SOLUTION

Wash and slice a fresh beet. Place about 4 slices of beet into a pan containing 1 cup of water. Heat until boiling and continue boiling for about 5 minutes. Remove the beet slices and allow the red liquid to cool. Store in dropper bottles. Beet juice is red in acidic solutions and blue in basic solutions.

BROMTHYMOL BLUE SOLUTION

Dissolve about 0.1 g (a pinch) of bromthymol blue powder in 1 L (1 qt) of distilled water. Stir well to dissolve. If the solution is green, it is neutral. Add a 4 percent solution of sodium hydroxide drop-by-drop until the color changes to deep blue. Store in dropper bottles. Bromthymel blue solution (BTB) is blue in basic solutions and yellow in acidic solutions.

Caution: BTB stains hands and clothing. Have students wear safety goggles when using BTB. Sodium hydroxide is corrosive. If spilled, flush affected areas with water.

IODINE STAINING SOLUTION

Iodine staining solution can be prepared by diluting tincture of iodine, purchased from a drugstore, with water. Add 50 mL (about ¼ cup) of water. Store the solution in brown dropper bottles.

Caution: Have students wear safety goggles when using iodine. Iodine stains hands and clothing. Iodine is toxic. Do not ingest.

INDOPHENOL BLUE INDICATOR

This vitamin C indicator can be prepared by dissolving 0.1 g of indophenol blue powder in 1 L (1 qt) of distilled water.

LIMEWATER

Fill a one-quart jar with water. Add 1 tablespoon of lime (used in making pickles) and stir. Fasten the lid and allow the solution to stand overnight. Pour off the clear liquid into a second jar. Be careful not to pour any of the lime that has settled to the bottom of the jar. Keep the jar of clear limewater tightly closed. Limewater is used to test for the presence of carbon dioxide.

CABBAGE INDICATOR

Tear 5 leaves of red cabbage into small pieces. Place the cabbage pieces in a small pan. Add 4 cups of hot water. Let the leaves soak for about ½ hr until the water is deep purple and cooled to room temperature. Strain this liquid into a storage bottle. Cover and store in the refrigerator. Cabbage indicator can be used to test for acids and bases.

CABBAGE PAPER

Pour 1 cup of cabbage indicator (above) into a bowl. Dip one coffee filter into the indicator. Place the wet paper on a cookie sheet. Continue wetting filter papers until the cookie sheet is filled with papers. Allow the papers to dry. The papers will be pale blue. Cut the dry papers into strips about 1.25 cm by 7.5 cm (0.5 inches by 3 inches). Store the strips in a ziplock plastic bag. Cabbage paper turns green in the presence of bases and pink to red in the presence of acids.

PHENOLPHTHALEIN SOLUTION

Purchase any laxative that contains phenolphthalein. With the back of a spoon, mash 4 tablets in a saucer. Pour the powder into a small cup. Add about 10 mL of rubbing alcohol. Let this mixture stand for about 15 minutes. Pour off the liquid and store in a dropper bottle. Phenolphthalein is purple in very basic solutions and colorless in acidic solutions.

TURMERIC INDICATOR SOLUTION

Obtain a package of turmeric from the spice section of the grocery store. Add ¼ teaspoon of turmeric to 4 tablespoons of rubbing alcohol. Stir to mix. Store in a dropper bottle. Turmeric solution stays yellow in the presence of acids and changes to purplebrown in the presence of bases.

Turmeric solution can also be made into indicator paper (see Cabbage Paper). Dry turmeric paper is bright yellow and changes to red in the presence of bases.

SIMULATED OCEAN SALT WATER

The salt content of ocean water varies from 2.5 percent to 3.5 percent. You can make salt water by mixing 100 g of table salt or sea salt in 3 L (3 qt) of water.

STARCH-IODINE INDICATOR

Mix 2 g of cornstarch with about 200 mL of water. Boil this mixture in a ceramic container to dissolve the cornstarch. Add 8 mL (1 ½ teaspoons) of this solution to 1 L (1 qt) of water. Add 1 mL (20 drops) of tincture of iodine. The color of this indicator should be royal blue. Starch-iodine indicator is used to test for vitamin C. In the presence of vitamin C the indicator becomes clear.

Caution: Iodine is toxic. Do not ingest. Iodine stains hands and clothing. Have students wear safety goggles when using iodine.

Fun Recipes

THE GREEN BLOB

Mix 2 common liquids to produce a green, jellylike blob of material. Begin by filling one half of a small baby food jar with steel wool. Add enough vinegar to cover the steel wool. Secure the lid and label the jar "Iron Acetate." Allow the jar to stand undisturbed for 5 days. Pour 1 tablespoon of the liquid Iron Acetate into a second jar. Add one tablespoon of household ammonia and stir. A dark green, jellylike substance forms immediately.

POLYMER

Prepare the following solutions ahead of time.
Solution A: Place about 1 L (1 qt) of water in a pan. Slowly sprinkle 40 g of polyvinyl alcohol on the water while stirring. Heat the mixture to about 90°C. Do not boil. Add several drops of blue food coloring. Stir the mixture for about 20 minutes until it looks like white corn syrup. Cool and store in a plastic jar.

Solution B: Dissolve 8 g of borax in about 200 mL of water. Add several drops of yellow food coloring. Stir until borax is dissolved. Store in a plastic jar.

To prepare the polymer: Place 30 mL of solution A in a paper cup. Add to this 10 mL of solution B. Stir with an ice-cream stick until a soft ball is formed. Remove the polymer ball from the cup and knead for about 5 minutes.

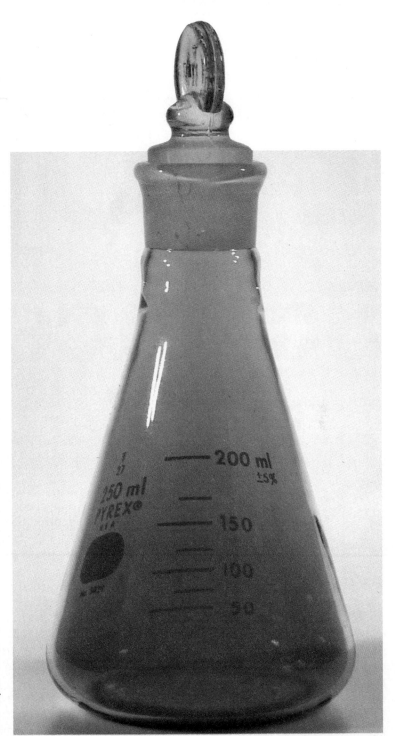

Substitutions

Iron filings can be made from a dry, fine-grade (grade 000) steel wool pad, the kind without soap. Put on gloves. Pull the steel wool pad into 2 halves. Hold the halves over a piece of paper and rub them together so that the iron filings fall on the paper. Carefully pour the iron filings into a dry jar. Cover until ready to use. ■

Project

Science for All

Science for All Americans is a report issued by the National Council on Science and Technology, a distinguished group of scientists and educators appointed by the American Association for the Advancement of Science (AAAS). This report consists of a series of recommendations concerning the scientific literacy of the American people. Science for All Americans details the understandings of science content and the habits of mind that are essential for all citizens in a scientifically literate society.

Scientific literacy—which embraces science, mathematics, and technology—has emerged as a central goal of education. However, studies have shown that scientific literacy eludes much of the population of the United States. A cascade of recent studies has made it abundantly clear that compared with national and world standards, education in the United States is failing too many students—and hence is not serving the nation's needs. By all accounts, the United States has no more urgent priority

than the reform of education in science, mathematics, and technology.

Reform is needed because this nation has not acted decisively enough to prepare young people, especially minority students, for a world that continues to change radically in response to the rapid growth of scientific knowledge and technological power. However, educational reform can not occur simply as a result of enacted legislation. It will take time, as well as determination, collaboration, the commitment of financial resources, and leadership to achieve educational reform.

But most of all, it will take a shared national vision of what Americans want their schools to achieve. Science for All Americans, part of the AAAS initiative called Project 2061, is intended to help formulate that vision.

In preparing its recommendations, the national council relied on reports of five independent panels of scientists. Additionally, the national council sought the advice of a large and diverse group of consultants and reviewers—scientists, engineers, mathematicians, historians, and educators. In all, the process took more than three years to complete, involved

hundreds of individuals, and culminated in the unanimous approval of Science for All Americans by the board of directors of the AAAS.

Project 2061 and Scientific Literacy

Science for All Americans, the first of a series of reports to be issued by Project 2061, says little about what ails the educational system. It points no blame and prescribes no specific remedies. Rather, its basic purpose is to characterize scientific literacy. Thus, its recommendations are presented in the form of basic learning goals for all American children. A fundamental premise of Project 2061 is that the schools do not need to be asked to teach more, they need to teach less and teach it better. Accordingly, the recommendations in Science for All Americans for a common core of learning are limited to the ideas and skills that have the greatest scientific and educational significance.

Project 2061 was begun in 1985, the year when Comet Halley approached the Earth. That recurrent phenomena inspired the project's name, for it was realized that the

2061:
Americans

children who were starting their school years then, would most probably be alive when the return of the comet to the Earth's vicinity occurred in 2061.

Project 2061 consists of a three-phase plan of purposeful and sustained action that will contribute to the critically needed reform in education in science, mathematics, and technology.

Phase I focused on the substance of scientific literacy. *Science for All Americans* and the other reports of the scientific panels are the chief products of this phase. The purpose of Phase I was to establish a conceptual base for reform by spelling out the knowledge, skills, and attitudes all students should acquire as a consequence of their K–12 school experience.

Phase II involves the transforma-

tion of recommendations made in *Science for All Americans* into several alternative curriculum models for use in various school districts and states. During this phase of the project, blueprints for reform relating to the education of teachers, the materials and technologies needed for teaching, the testing and organization of schooling, as well as educational policies and educational research are to be developed. While engaged in creating these resources, Project 2061 is also trying to enlarge significantly the nation's pool of experts in science curriculum reform. Project 2061 will continue in its efforts to publicize the need for nationwide scientific literacy.

Phase III will consist of a widespread collaborative effort that will last a decade or longer. During this final phase, many groups active in educational reform—those who will take into account the history, economics, and politics of change—will use the resources presented in Phases I and II to move the nation toward its goal of producing a scientifically literate society by the year 2061. Strategies for implementing reform of education in science, mathematics, and technology in the na-

tion's schools will be developed by those with a stake in the effectiveness of the schools. ■

RECOMMENDATIONS

Following is a brief summary, in general terms, of the national council's recommendations:

- Being familiar with the natural world and recognizing both its unity and diversity.

- Understanding the key concepts and principles of science.

- Being aware of some of the important ways in which science, mathematics, and technology depend upon one another.

- Knowing that science, mathematics, and technology are human enterprises and knowing what that implies about their strengths as well as their limitations.

- Having a capacity for scientific ways of thinking.

- Using scientific knowledge and ways of thinking for individual and social purposes.

What Is Project Scope, Sequence, and Coordination?

The goal of Project Scope, Sequence, and Coordination (SS&C), initiated by the National Science Teachers Association, is a major reformation of the teaching of science at the secondary level. SS&C recommends that all students study science every year for six years. It advocates carefully sequenced, well-coordinated instruction in all the sciences. As opposed to the traditional "layer-cake" curriculum in which science is taught in discrete, year-long disciplines, the NSTA project calls for the study of each discipline to be spread out over several years.

SS&C proposes that learning be developed through hands-on experience first, with terminology, symbols, and equations to follow later. It also proposes that a major focus of science teaching should be the introduction of fewer topics designed to encourage a greater depth of understanding. Students will develop skills to help them answer questions such as

- How do we know?
- Why do we believe?
- How do we find out?
- What is the evidence?
- What does it mean?

Why Is Reform Needed?

Today the perception exists that schools are failing to prepare students for life in a world that depends more and more on sophisticated and rapidly changing science and technology. The majority of students leave school, whether dropouts or graduates, without a basic understanding of science, mathematics, or technology.

It is suggested that students perceive textbook-driven science courses to be difficult and boring. They feel that currently structured science courses have little relevance to their lives. In fact, more than half of our students do not take another science course after the tenth grade.

Student Assessment

Because the SS&C reform effort presents such a radical departure from the normal patterns of instruction at the secondary level, a radical change is also necessary to assess the progress of students taking part in the pilot programs. With funding provided by the United States Department of Education, NSTA is presently developing new methods of student assessment linked to performance. This new method of assessment is based on technology that uses interactive videodiscs. Student assessment will require students to demonstrate *why* they hold a particular belief, and *how* they know that something is correct. They will also be asked to demonstrate a knowledge of terms, using real objects and phenomena.

What Are the Elements of SS&C?

SCOPE

The fundamental goal of the Scope, Sequence, and Coordination Project is to make the basic sciences understandable and enjoyable for all students. To accomplish this goal, the SS&C project will concentrate on providing opportunities for students to engage in direct experience with the natural world, before they are exposed to science terms, to symbols that represent physical quantities, or to the equations for relationships that scientists use to explain the natural world.

In the restructured program, the number of science topics and the terminology will be reduced greatly. Fewer topics sequenced and taught over several years will produce greater understanding of how science can be used to approach everyday issues that have scientific or technical components.

SEQUENCE

SS&C reform emphasizes the appropriate sequencing of instruction, which takes into account how students learn. Concepts, principles, and laws of science can be approached at successively higher levels of abstraction over several years. This sequencing should make it possible for all students to learn.

The practical application of science that focuses on problems and issues of personal concern to students should begin in the first year of science instruction. Later, in the upper grades, applications should become more "global." Treated in this way students will be better able to relate science to themselves and their lives.

COORDINATION

Biology, chemistry, earth/space science, and physics share certain topic areas. Coordination among these four subjects leads to an awareness of the interdependence of the sciences and shows students how the sciences fit together as part of a larger body of knowledge. In the middle grades, the sciences will be integrated and taught by one or two teachers. At the high school level, the sciences will be taught as separate subjects, while still retaining strong coordination.

What Are the Advantages of SS&C?

SS&C

- Takes advantage of spaced learning theories.
- Provides experience with science phenomena before science terminology is introduced to students.
- Builds the concepts of science through repeated experiences in differing contexts.

- Assesses the depth of student understanding, not only a knowledge of facts or discrete bits of information.
- Applies science knowledge to personal and societal problems.

What Are the Anticipated Outcomes of SS&C?

It is expected that SS&C will

- Develop science literacy in citizens of the United States.
- Increase the number of students studying science at advanced levels.
- Encourage females and members of all minority groups to study science.
- Create a greater level of understanding of scientific content materials.
- Lead to development of new kinds of textbooks and instructional materials.
- Provide more effective ways to assess student learning.

Supporting References

Education That Works: An Action Plan for the Education of Minorities. The Action Council for Minority Education. Cambridge, Massachusetts: Massachusetts Institute of Technology. January 1990.

Science for All Americans: Project 2061. American Association for the Advancement of Science. Washington, DC.

Dempster, F. "The Spacing Effect: A Case Study in the Failure to Apply the Results of Psychological Research." *American Psychologist*, 43, 1988. 627–634.

Piaget, Jean. *To Understand Is to Invent: The Future of Education*. 1973. New York: Grossman Publishers.

Resnick, Lauren B. "Mathematics and Science Learning: A New Conception." *Science*, 220, 1983. 477–478. ■

TO THE TEACHER

The ability to design and draw graphs is one of the most important skills students must learn. Very often it is one of the most difficult concepts to present, particularly in a format applicable to all ability levels. The following booklet—which should be reproduced and handed out to each student—will enable you to teach graphing concepts in an entertaining and pedagogically sound manner. The classroom-tested materials begin with simple graphing skills and proceed to more difficult graphing skills.

The booklet itself is a reader. Following the booklet are 11 activities keyed to the booklet. These activities are called Brain Stretchers. After the Brain Stretchers is an answer key. It is suggested that you begin this graphing skills material very early in the school year. Students can proceed at their own pace, or you may want to assign a section each week. All sections end with a note to go to a Brain Stretcher. After each section has been read, distribute a copy of the appropriate Brain Stretcher, to be done as either an in-class assignment or as homework.

graphing

BY

B.K. HIXSON

INTRODUCTION

Welcome to science, whippersnapper! Wilshere A. Smith here, and I have the distinct privilege of being your host as you study graphing. Actually I'm a cartoon character with a lot of personality, but don't let that fool you, I've got this science stuff down pat. I'd like to introduce you to Ophidia, a good friend of mine. She is generally a good helper, but as you'll find out, sometimes she can be quite a bother. So, Ophidia my friend, if you would be so kind as to get out of the way, we'll proceed.

Jargon: a specialized vocabulary used to describe things in a specific field of study.

As your host, I'm going to help you out two ways. First, I need to be sure that we can communicate with one another, so to do that you'll need to learn the scientific **jargon** (jar • gun) for this unit. Look at the left margin. That's where I'll define all of your vocabulary words, which are printed in boldface. Also, each main idea I want you to learn is found in the margin. And just to be sure you don't miss these babies, I'll put stars by them. There is a catch though—these ideas are asked as questions that you need to find the answer to. Once you figure out the answer, get that hot little pencil moving and fill in your answer right beside the question. This way you'll have your own study guide to help you study for tests. So, once again welcome to science. It's nice to have you here.

DATA

Data: information of observations.

Let's get down to the business at hand. **Data** is information. Pure and simple, data is information. Look at these examples:

- Magic Johnson's height

- The number of centimeters of snow that fell at Timberline Lodge every day for one month

- The number of kilometers from New York to Los Angeles

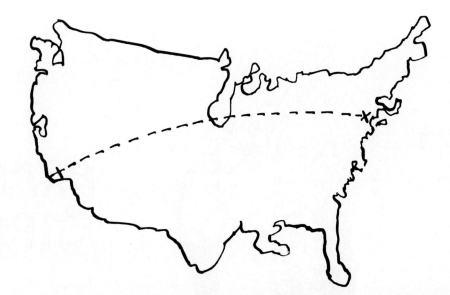

Each of these three items is a little bit of data.

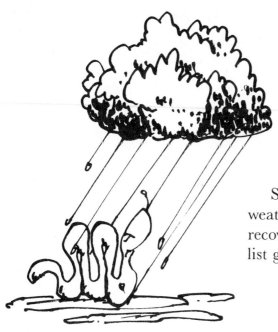

Scientists use data all the time. Data helps predict the weather, improve stereo equipment, and aid in the fast recovery of patients in hospitals. But it doesn't stop there, the list goes on and on.

How is data valuable to scientists?

As scientists, you'll need to be able to record information, evaluate experiments, and draw conclusions. To do all of this, you need to know how to use data tables and draw graphs, which is exactly what this unit is about.

BRAIN STRETCHER #1

DATA TABLES

If we collect just a little bit of data it's no problem to keep track of it. But what happens when we get a lot? Right, our organization improves real fast and to organize data we use a **data table.**

Data table: a way to organize data in columns so it is neat and readable.

Here is a data table constructed for an imaginary science experiment.

Heating of Compound X

Time (minutes)	0	1	2	3	4	5	6	7	8	9	10
Temperature (°C)	20	21	23	27	35	45	61	69	71	73	74

Title: a brief way to describe the content of a book, graph, or data table.

Data tables begin with a **title.** The title of this data table is "Heating of Compound X." The title tells the reader exactly what the data in the table refers to. Without a title, it's possible that the person studying the data table wouldn't be able to figure out what all the numbers meant. And that can be very frustrating. When you construct a data table always be sure to begin with a title that describes the data it contains.

Variable: a word used in a data table to describe what information is being collected.

A data table includes variables and units. A **variable** (vary • a • bull) describes *what* information you are recording. A **unit** tells *how* you are going to measure that variable.

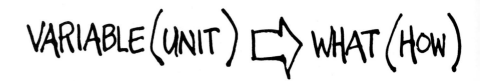

Unit: a word or symbol used in a data table that tells how the information was measured.

Heating of Compound X

Time (minutes)	0	1	2	3	4	5	6	7	8	9	10
Temperature (°C)	20	21	23	27	35	45	61	69	71	73	74

Why is it important to have variables and units included in a data table?

In this data table, the first column of data is the time variable, which is recorded in units of minutes. It's easy to figure out the units because they are always in parentheses. The second column is for the temperature variable measured in degrees Celsius. Think of variables and units in these terms:

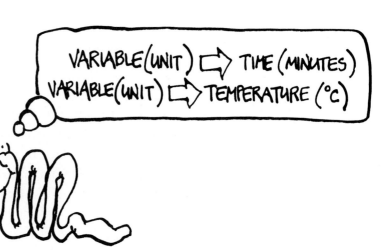

Ordered pairs:
two pieces of data directly corresponding to one another.

Data is organized in a data table. If you look carefully, you'll find the numbers are organized in groups called **ordered pairs.** They are called ordered pairs because the numbers always go together. It's easy to tell ordered pairs in a data table because they're piled on top of one another.

Look at the data table and you'll see that the combination of 2 minutes and 23 degrees Celsius (2 min, 23 °C) is one ordered pair.

Heating of Compound X

Time (min.)	0	1	2	3	4	5	6	7	8	9	10
Temperature (°C)	20	21	23	27	35	45	61	69	71	73	74

5 minutes and 45 degrees Celsius is another ordered pair (5 min, 45 °C).

There are 11 ordered pairs in this data table:

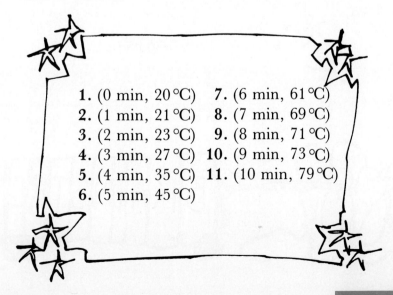

1. (0 min, 20 °C)   **7.** (6 min, 61 °C)

2. (1 min, 21 °C)   **8.** (7 min, 69 °C)

3. (2 min, 23 °C)   **9.** (8 min, 71 °C)

4. (3 min, 27 °C)   **10.** (9 min, 73 °C)

5. (4 min, 35 °C)   **11.** (10 min, 79 °C)

6. (5 min, 45 °C)

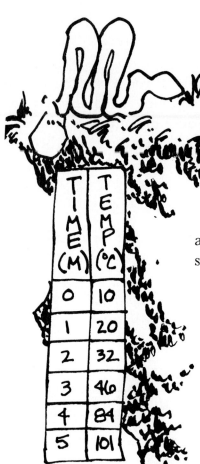

TIME (M)	TEMP (°C)
0	10
1	20
2	32
3	46
4	84
5	101

Now, if for some reason you find a data table that runs up and down instead of across, the ordered pairs are still there, just side by side now.

Heating of Compound X

Time (minutes)	Temperature (°C)
0	20
1	21
2	23
3	27
4	35
5	45
6	61
7	69
8	71
9	73
10	74

What is the best way to identify an ordered pair?

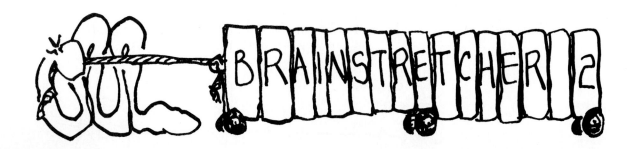

A COMPLETE DATA TABLE

A properly constructed data table follows these guidelines:

1. a descriptive title
2. variables describing what information has been collected
3. units telling how those variables were measured
4. data collected in ordered pairs
5. all work done neatly

Each time you construct a data table check it against these guidelines to be sure everything is there. Not only will you be getting great grades, but pretty soon you won't need the guidelines and you'll be producing perfect work without even trying.

BRAIN STRETCHER #5

"THE ONLY PLACE WHERE SUCCESS COMES BEFORE WORK IS IN THE DICTIONARY."

READING A GRAPH

Graph: a picture of information in a data table.

By now you're an expert at reading and constructing data tables. Now it's time to master reading a **graph.** A graph is an exact picture of information in a data table. It's usually a line that shows a relationship between all of the numbers in a data table. The following data table and graph are for the Heating of Compound X.

Heating of Compound X

Time (minutes)	0	1	2	3	4	5	6	7	8	9	10
Temperature (°C)	20	21	23	27	35	45	61	69	71	73	74

What are the similarities between a data table and a graph?

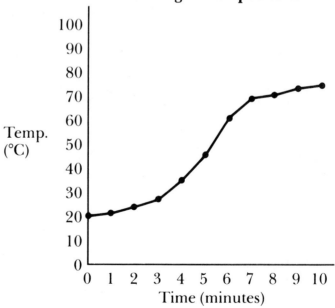

Heating of Compound X

Temp. (°C)

Time (minutes)

Here's the kicker, if a graph is an exact picture of the data table, then everything that's in the data table has to be in the graph—everything!! Check it out.

1. The same descriptive title—The Heating of Compound X.
2. The same variables and units—time (min) and temp. ($°C$). One on each side.
3. The same data.

A GRAPH IS A PICTURE OF A DATA TABLE

Horizontal axis: the axis that goes across the bottom of the graph

Vertical axis: the axis that runs up and down on the side of the graph.

The secret is knowing what to do with the ordered pairs. Take the second ordered pair (1 min, 21°C) and split them up. Place the 1 minute along the **horizontal axis** labeled "Time" at the 1 minute mark. Put the 21°C on the **vertical axis** labeled "Temperature" at the 21°C mark. Now follow each line into the middle until they both meet.

Intersection: the crossing of two lines when graphing.

The point where these two lines cross is called the **intersection** of the lines. And the point showing the location of that intersection is called a **data point.**

Data point: the place where the two data lines cross (or intersect).

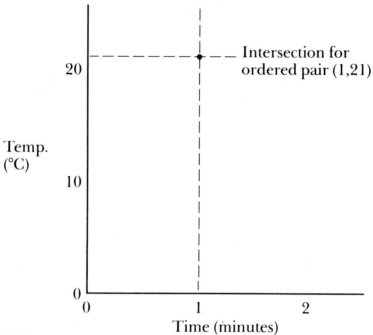

(1 min, 21°C)

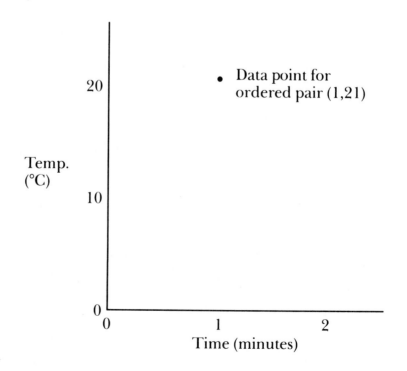

How do you find the ordered pair given only the data point?

To read a graph find a data point. Simply trace the horizontal and vertical lines back to each axis and you've found the ordered pair for that data point.

Plotting: finding the data point for an ordered pair.

Locating the data point by finding the intersection of the two lines is called **plotting.** Once all the data points are plotted and a line drawn to connect them, you have a simple way to "read" all the data in a data table quickly.

PLOTTING A LINE GRAPH

In order to plot a line graph, you'll need the materials listed below (be sure your brain is plugged in, too). We're going to use Heating Compound X for our example.

1. Graph paper (naturally)
2. Ruler
3. Pencil with an eraser (pens make permanent mistakes)

You'll also need some guidelines:

1. Always use graph paper when constructing a graph. It's easier to plot and more accurate than notebook paper.
2. Draw all lines with a ruler. It gives the graph a professional appearance and reduces the chance of error.
3. When drawing the horizontal and vertical axes start no less than three squares from the bottom of the paper and three squares from the left side. This gives you enough margin to label the axes legibly.
4. Do all your work in pencil. Pens don't erase, and neatness *does* count.
5. A complete graph has a descriptive title, both axes labeled correctly, data points plotted neatly, and a line connecting data points drawn with a ruler.
6. Use the whole piece of graph paper to do a graph. It is easier to read than a minigraph.

PAPER, PENCIL, RULER, CHECK.

If you follow these simple guidelines then making a graph is easy. First, draw the horizontal and vertical axes using your ruler. Be sure to indent three spaces.

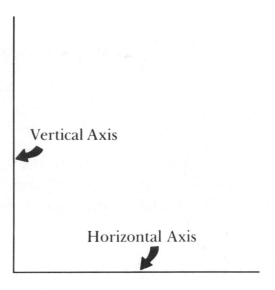

Next, look at the data table and decide which variable is the most consistent. In our example it's time. *The most consistent variable is labeled on the horizontal axis.* Label the vertical axis with the other variable.

Temp. (°C)

Time (minutes)

Intervals: an even spacing of the numbers along the axis of the graph.

Third, number each axis in even **intervals.** Intervals are determined by looking at the smallest and largest numbers in the data table. Devise an even way to space them along the axis. Here's our data table.

Heating of Compound X

Time (minutes)	0	1	2	3	4	5	6	7	8	9	10
Temperature (°C)	20	21	23	27	35	45	61	69	71	73	74

For time, the smallest number is 0 and the largest is 10. The numbers between are in even intervals of 1's. Because the work is already done, label the horizontal axis straight off the data table.

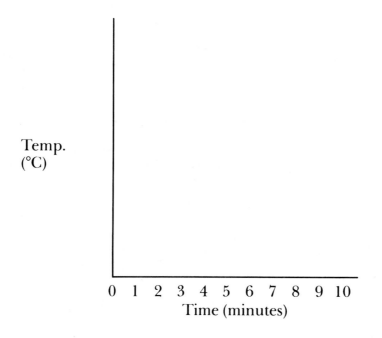

Temp. (°C)

0 1 2 3 4 5 6 7 8 9 10
Time (minutes)

INTERVALS, AXIS, DATA, SCHMATA, I'M GONNA EAT MORE FLIES.

To determine the labels for the vertical axis we need to think a bit. The smallest number is 20, the largest number is 74. The rest of the numbers between are hardly what you would call even. If we label the axis by 10s and then plot the data points, we'll get a good picture of the data. You will most always satisfy your needs if you label the axis in intervals of 5, 10, or 100.

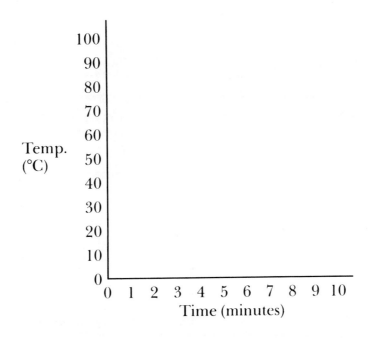

Once the axes are drawn and labeled it's time to plot the data points. Using the data table below, split up the ordered pairs. Find each half of the pair on the respective axis, and follow the lines until they intersect. You can see how all of the ordered pairs should be plotted.

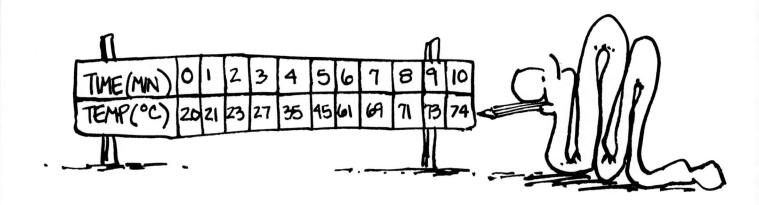

TIME (MIN)	0	1	2	3	4	5	6	7	8	9	10
TEMP (°C)	20	21	23	27	35	45	61	69	71	73	74

Heating of Compound X

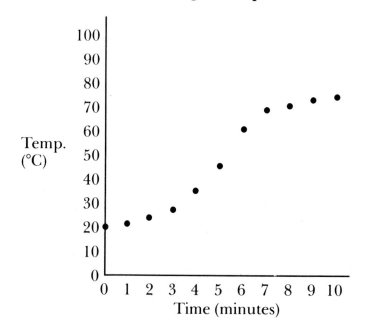

Line graph: a graph drawn using data points.

After all the data points are plotted, get out that hot little pencil. Line up the data points with a ruler, and connect the dots. When all the points are connected, you'll have a **line graph** like the one below.

Explain why an accurate relationship between the ordered pairs is shown only if even intervals are marked on the axes.

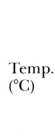

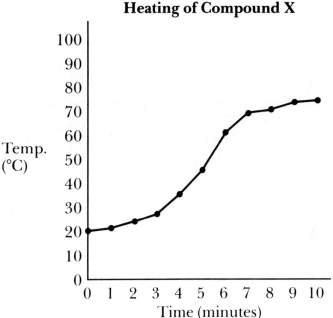

Be sure to copy the descriptive title from the top of the data table onto the top of the graph.

BRAIN STRETCHER 7

READING MULTIPLE LINE GRAPHS

When there is just one line on the graph it is called a single line graph. That makes sense. But, if there are two or more lines on a single graph it's called a multiple line graph. The data table below can be used to produce one of these graphs. Multiple line graphs are especially helpful when comparing several different experiments utilizing similar data.

Heating of Compounds X, Y, Z

Time (minutes)	0	1	2	3	4	5	6	7	8	9	10
Temp. X (°C)	20	21	23	27	35	45	61	69	71	73	74
Temp. Y (°C)	25	30	32	33	33	35	40	60	70	80	83
Temp. Z (°C)	28	35	55	56	57	57	69	75	78	78	79

There are still only two variables; time and temperature. So, the graph is still constructed the same as the first example. The tricky part comes in figuring out the ordered pairs because there are three temperature columns and only one time column. The answer is to recycle the time variable by matching it to the temperatures directly below it each time.

ORDERED PAIRS

X LINE (0, 20°)
Y LINE (0, 25°)
Z LINE (0, 28°)

BRAIN STRETCHER 8

CONSTRUCTING A MULTIPLE LINE GRAPH

The easiest way to construct a multiple line graph is to plot one line at a time. Using this method reduces the chance of error. Here's how it works, step-by-step. First plot the ordered pairs for time and temperature X. We'll call the line you produce line X.

Heating of Compounds X, Y, Z

Time (minutes)	0	1	2	3	4	5	6	7	8	9	10
Temp. X (°C)	20	21	23	27	35	45	61	69	71	73	74
Temp. Y (°C)	25	30	32	33	33	35	40	60	70	80	83
Temp. Z (°C)	28	35	55	56	57	57	69	75	78	78	79

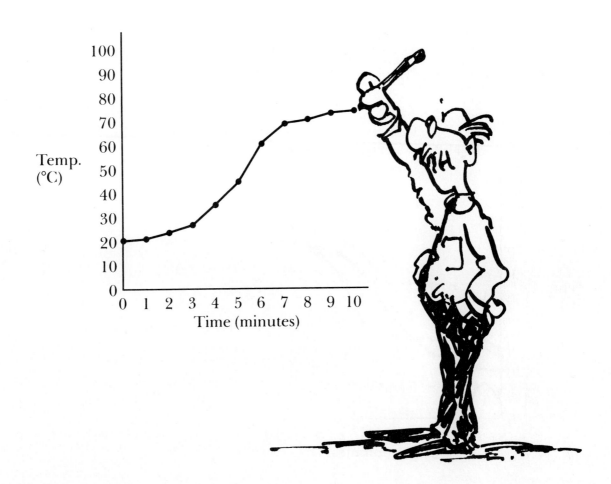

Now that line X is complete move on to line Y. Plot line Y on the same graph.

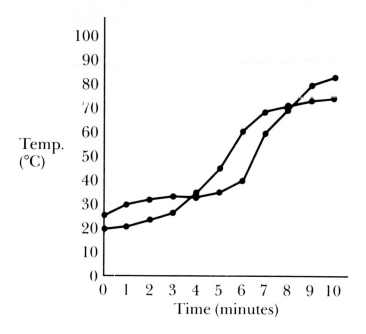

Lines X and Y are on the graph. As soon as line Z is plotted the graph will be finished.

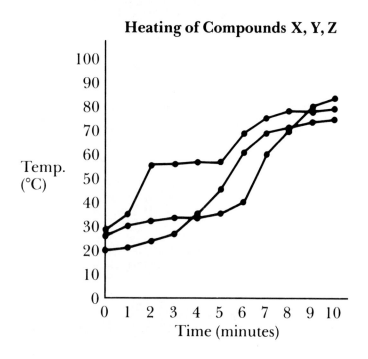

Heating of Compounds X, Y, Z

What are the advantages to using a multiple line graph?

And there you have it, a finished line graph. For the sake of clarity, you may wish to use a different color for each line or different symbols.

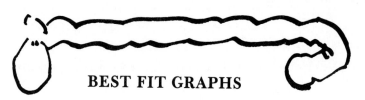

BEST FIT GRAPHS

Best fit graph:
a graph with a line passing through many but not all plotted points. Best fit graphs allow scientists to predict various unplotted points on a graph.

A line graph is a very precise way of recording data. Sometimes though, scientists prefer to find an average or would like to be able to predict what might happen. To do this they use a **best fit graph.**

A best fit graph is a guesstimate, somewhere between a guess and an estimate, or an educated guess. You construct the graph and plot the data exactly as you would for a line graph. The difference is the way you draw the line. You can see that the best fit line in the graph below is smooth and continuous. The line flows through most of the data, but doesn't necessarily have to touch all of the data points.

Heating of Compound X

Time (minutes)	0	1	2	3	4	5	6	7	8	9	10
Temperature (°C)	20	21	23	27	35	45	61	69	71	73	74

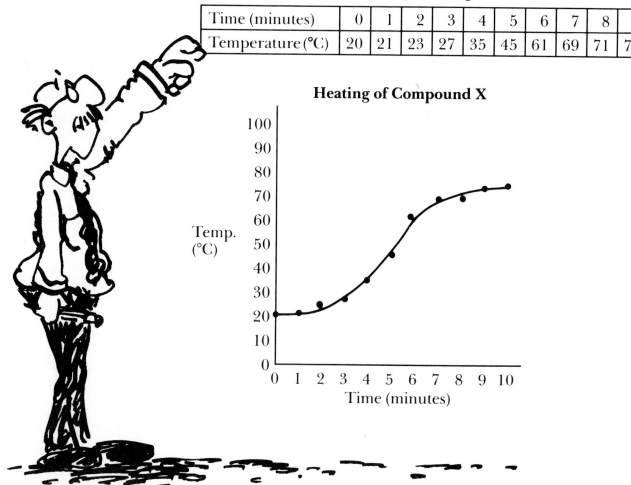

Heating of Compound X

The best fit line goes through as many of the data points as possible, but remains a smooth curve. The advantage to this kind of graph is that it allows us to predict data that we did not collect. The prediction may not be exact, but it will be very close. For example, what would be the temperature after 4 1/2 minutes of heating? We didn't directly measure the temperature at that point, but we can look at the graph and make a good guess at the probable temperature.

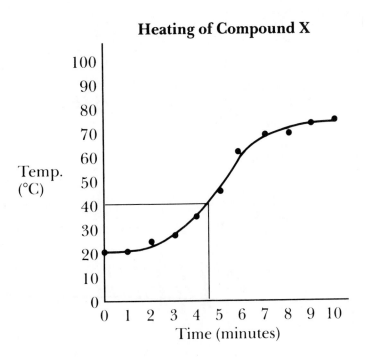

The answer to our question is probably close to 40 °C.

What are the differences between a line graph and a best fit graph?

Best fit graphs are valuable for two reasons:

1. Best fit graphs allow us to predict information by showing a representative curve of the data collected.
2. Best fit graphs save time because they can be plotted using a small sample of data.

THE END

BRAIN STRETCHERS 10 & 11

Name ______________________________ Class ____________________ Date ________________

BRAIN STRETCHER #1

Directions: Flip a penny 5, 10, 15, 20, 25, 30, and 35 times separately. Record the number of times the coin lands heads up for each test. Organize your data here below. There is no right or wrong way to organize your data. Just make sure you record it accurately.

Vocabulary: Define the following words.

1. Jargon—

2. Data—

Question: Answer the following question.
How is data valuable to scientists?

____

____

Name _____ Class _____ Date _____

Directions: Identify the different parts of the data table by answering the questions that follow it.

High Daily Temperature

Time (days)	1	2	3	4	5	6	7	8	9	10	11	12	13	14	15
Temperature (°C)	27	28	31	30	34	38	42	18	20	24	21	29	33	28	38

1. The title is _____ .

2. The two variables are _____ and _____ .

3. The unit for time is _____ and for temperature is __

4. List the ordered pairs.

_____ _____ _____

_____ _____ _____

_____ _____ _____

_____ _____ _____

_____ _____ _____

5. The highest daily temperature was _____, which occurred on day _____ .

6. Day 10 had a temperature of _____ .

7. On two days, the temperature recorded was 28°C. One was day _____ and the other was day _____ .

Vocabulary: Define the following words.

1. Data table—

2. Title—

3. Variable—

4. Unit—

5. Ordered pair—

Questions: Answer the following questions.
Why is it important to have variables and units included in a data table?

What is the best way to identify an ordered pair?

Name ____________________ Class ____________________ Date ____________________

BRAIN STRETCHER #5

Directions: Using the guidelines in your graphing booklet, correct each of the following data tables.

1.

Height of Corn Plants

Time	1	2	3	4	5	6	7	8
Height	5	25	70	82	120	306	420	570

2.

Time (weeks)	1	2	3	4	5	6	7	8
Height (cm)	5	25	70	82	120	306	420	570

3.

Time	1	2	3	4	5	6	7	8
Height	5	25	70	82	120	306	420	570

4.

Height of Corn Plants

Height (cm)								

5.

Height of Corn Plants

(weeks)	1	2	3	4	5	6	7	8
(cm)	5	25	70	82	120	306	420	570

brain stretcher 4

Directions: Using the graph, answer the questions that follow it.

Cooling of Substance Z

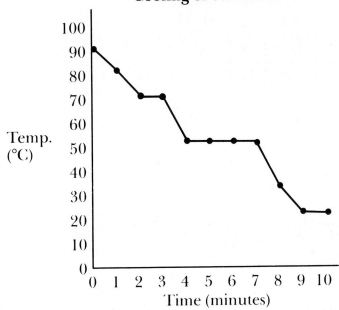

1. The title of the graph is _____

2. The two variables are _____ and _____

3. The unit for time is _____ and for temperature is _____
   _____ .

4. At ___3___ minutes it was _____ °C.

5. At ___9___ minutes it was _____ °C.

6. At _70°C_ the time was _____ minutes.

7. At _30°C_ the time was _____ minutes.

8. At what temperature did substance Z level off for 4 minutes?
   _____

9. What was the starting temperature of substance Z? _____

10. The total temperature loss for substance Z was _____

Vocabulary: Define the following words.

1. Graph—

2. Horizontal axis—

3. Vertical axis—

4. Intersection—

5. Data point—

6. Plot—

Question: Answer the following question.
What are the similarities between a data table and a graph?

brain stretcher 5

Directions: Using the following graph, construct a data table in the space below. Remember the guidelines in the graphing booklet.

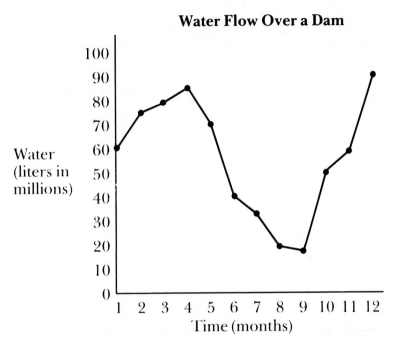

Water Flow Over a Dam

Question: Answer the following question.
How do you find an ordered pair given only the data point?

Directions: Compare the data table with the two graphs and answer the questions.

Average Daily Temperatures

Time (days)	1	2	3	4	5	6	7	8	9	10
Temperature (°C)	10	15	20	30	40	40	35	25	15	5

Average Daily Temperatures

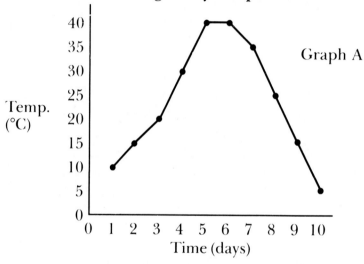

Average Daily Temperatures

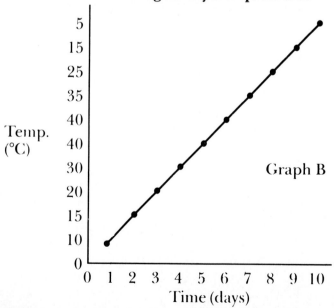

1. What does the line in graph A show the temperature doing over 10 days?

2. What does the line in graph B show the temperature doing over 10 days?

3. What is the major difference in the lines of the two graphs?

4. Look at the data for temperature in the data table. Describe what happens to the temperature.

5. Which graph shows an accurate picture of the data?

6. How would you change the graph that does not show an accurate picture of the data?

BRAIN STRETCHER 7

Directions: Using the data tables below, construct a line graph for each. Use a separate sheet of graph paper for each graph.

Average Rainfall in Willamette Valley

Time (months)	1	2	3	4	5	6	7	8	9	10	11	12
Rainfall (mL)	15	21	28	24	16	8	2	1	2	3	5	10

High Daily Temperatures

Time (days)	1	2	3	4	5	6	7	8	9	10	11	12
Temperature (°C)	27	28	31	30	33	37	36	18	20	21	24	30

BRAIN STRETCHER 8

Directions: Using this graph, answer the following questions.

Growth of Three Pea Plants

1. For the weeks given, record the heights of the three plants.

Week	Plant A	Plant B	Plant C
3	_____	_____	_____
7	_____	_____	_____
10	_____	_____	_____

2. How many weeks did it take for plant A to reach 90 cm? _____.

3. Which plant remained at about 30 cm for 4 weeks? _____.

4. Which plant remained at 90 cm for 2 weeks? _____.

5. When plant A was 50 cm tall how tall was plant C? _____.

6. Which plant grew the most? _____.

7. Which plant grew the least? _____

8. Which plant was probably sick?

Directions: Construct a multiple line graph for each of the two data tables below. Use a separate piece of graph paper for each graph.

Gain in Mass of Young Mice

Time (weeks)	1	2	3	4	5	6	7	8	9	10
Mass 1 (mg)	25	115	140	210	400	720	780	810	830	900
Mass 2 (mg)	20	120	125	200	250	300	400	520	540	570
Mass 3 (mg)	30	140	160	220	380	680	690	770	800	950

High Temperatures For Three Cities

Time (days)	1	2	3	4	5	6	7	8	9	10
Temp. (LA) (°C)	28	29	22	27	26	27	29	31	30	24
Temp. (SF) (°C)	18	22	20	17	21	21	20	18	21	19
Temp. (PD) (°C)	11	10	14	13	15	16	15	11	15	17

Question: Answer the following question.
What are the advantages of a multiple line graph?

brain stretcher 10

Directions: Practice making a smooth curve on the following graphs.

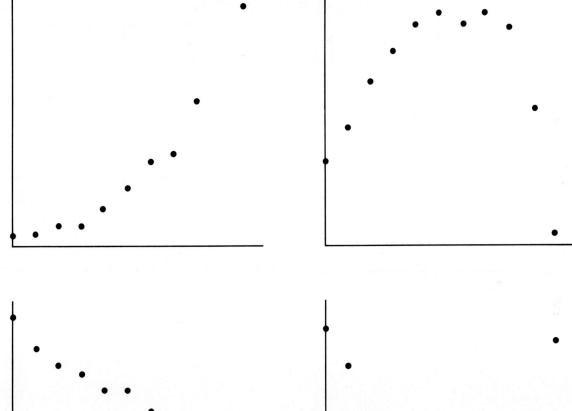

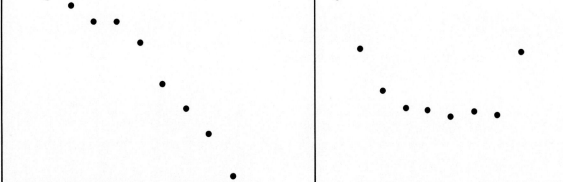

Question: Answer the following question.
What are the differences between a line and best fit graph?

BRAIN STRETCHER #17

Directions: Construct a best fit graph for each of the two data tables below. Use a separate piece of graph paper for each graph.

Height of Corn Plants

Time (weeks)	1	2	3	4	5	6	7	8
Height (cm)	5	25	70	82	120	306	420	570

High Daily Temperature

Time (days)	1	2	3	4	5	6	7	8	9	10	11	12	13	14
Temperature (°C)	27	28	31	30	34	37	37	18	20	24	21	29	32	37

Answer Key

BRAIN STRETCHER 1

Students should organize data in a data table. Ordered pairs will show number of flips versus number of heads.

Vocabulary   1 Jargon—A specialized vocabulary used to describe things in a specific field of study   2. Data—Information based on observation

Question   Answers will vary.

BRAIN STRETCHER 2

1. High Daily Temperature   2. Time; Temperature
3. days; °C
4.

1 27	6 38	11 21
2 28	7 42	12 29
3 31	8 18	13 33
4 30	9 20	14 28
5 34	10 24	15 38

5. 42°C; 7   6. 24°C   7. 2; 14

Vocabulary   1. Data table—A method of organizing data in columns   2. Title—A concise way to describe the content of a graph or data table among other things 3. Variable—A word used in a data table that describes the information being collected   4. Unit—A word or symbol in a data table that tells how the information was collected 5. Ordered pair—Two pieces of data directly corresponding to one another

Questions   A variable describes what information is being studied. A unit tells the system of measurement used to obtain each variable.   An ordered pair is a group of two members, each member corresponding to a variable.

BRAIN STRETCHER 3

1. Lacks units: (weeks); (cm)   2. Lacks title: Height of Corn Plants   3. Lacks title and units: Height of Corn Plants; (weeks); (cm)   4. Lacks first variable and unit—Time (weeks)—and all data:

1	2	3	4	5	6	7	8
5	25	70	82	120	306	420	570

5. Lacks variables: Time; Height

Data table

Heads from Flipping Coins

Number of Flips	5	10	15	20	25	30
Number of Heads			Answers will vary.			

BRAIN STRETCHER 4

1. Cooling of Substance Z   2. Temperature; Time 3. minutes; °C   4. 70   5. 20   6. 2 and 3   7. 8 8. 50°C   9. 90°C   10. 90°C – 20°C = 70°C

Vocabulary   1. Graph—A picture of information in a data table   2. Horizontal axis—The axis (line) that runs across the bottom of a graph   3. Vertical axis—The axis (line) that runs up and down the side of a graph   4. Intersection—The crossing of two lines on a graph   5. Data point—The place where two lines on a graph cross (intersect)   6. Plot—A method of finding the data point for an ordered pair

Question   A graph is an exact picture of information in a data table.

BRAIN STRETCHER 5

Water Flow Over a Dam

Time (months)	1	2	3	4	5	6	7	8	9	10	11	12
Water* (liters in millions)	60	75	79	85	70	40	32	19	17	50	59	90

Student approximations should fall within plus or minus two of these numbers.

Question   Trace a horizontal and vertical line from the data point to each axis to find the ordered pair.

BRAIN STRETCHER 6

1. Goes from a low of 10°C to a high of 40°C between days 6 and 7; then falls to a low of 4°C at day 10.   2. Goes from a low of 10°C to a high of 40°C between days 6 and 7; then falls to a low of 4°C at day 10.   3. The vertical axis in graph A begins with the lowest temperature and is then broken down into increasing units. In graph B, the vertical axis units are arranged in the order they appear in the data table.   4. The temperature rises, then falls.   5. Graph A 6. Reorder the units in the vertical axis from 0°C to 40°C in equal increments so that the lowest temperature is at the bottom and the highest temperature is at the top.

BRAIN STRETCHER 7

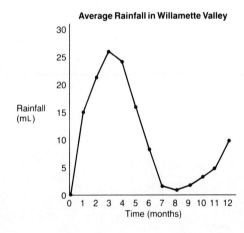

Average Rainfall in Willamette Valley

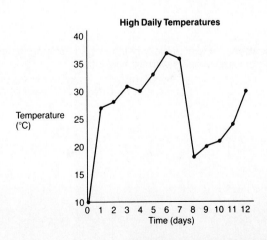

High Daily Temperatures

BRAIN STRETCHER 8

1.
Plant A	Plant B	Plant C
70 cm	40 cm	30 cm
90 cm	80 cm	38 cm
100 cm	97 cm	64 cm

2. 6   3. Plant C   4. Plant A   5. 30 cm   6. Plant A
7. Plant C   8. Although Plant C grew the least, it would be incorrect to assume from the data that it was sick.

BRAIN STRETCHER 9

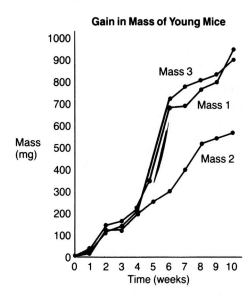

Gain in Mass of Young Mice

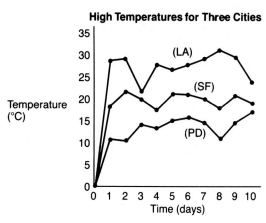

High Temperatures for Three Cities

Question   It is especially helpful when comparing several different experiments with similar data.

BRAIN STRETCHER 10

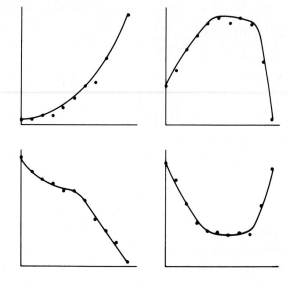

Question   A line graph has a line that passes through every point on the graph. A best fit graph is a smooth, continuous line that flows past most of the data points but doesn't necessarily touch them all.

BRAIN STRETCHER 11

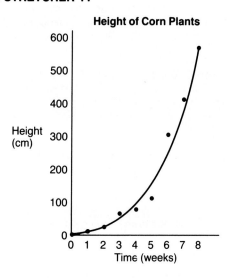

Height of Corn Plants

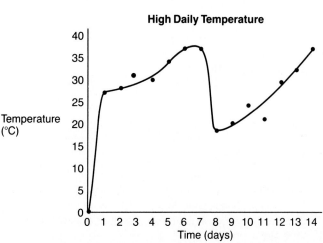

High Daily Temperature

Credits

Cover Photo: Ken Karp

3 center: Ken Lax for Silver Burdett & Ginn; **4** top: Will & Deni McIntyre/Photo Researchers, Inc.; bottom: Ken Karp/Omni-Photo Communications, Inc.; **54** Kevin Horan/Stock Boston, Inc.; **55** IBM Corporation; **56** Will & Deni McIntyre/Photo Researchers, Inc.; **57** Naoki Okamoto/Black Star; **58** top: Silver Burdett & Ginn; bottom: Martin Dohrn/Science Photo Library/Photo Researchers, Inc.; **59** left: NASA; right: R. Andrew Odum/Peter Arnold, Inc.; **60** Wilson North/International Stock Photography, Ltd.; **61** Hale Observatories; **62** J. Alex Langley/DPI; **63** Dr. Jan Lindberg; **64** Terry G. Murphy/Animals Animals/Earth Scenes; **65** Ken Karp/Omni-Photo Communications, Inc.; **66** top: Ken Karp; bottom: Barbara Kirk/Stock Market; **67** Ken Karp; **68** Richard Haynes for Silver Burdett & Ginn; **69** Michael Newman/Photoedit; **71** Silver Burdett & Ginn; **74** Diane Graham-Henry/Tony Stone Worldwide/Chicago Ltd.; **77** Ken Lax for Silver Burdett & Ginn; **78** Phil Degginger; **80** Ken Karp; **81** Bob Daemmrich/Stock Boston, Inc.; **82** Illustration by John Tenniel from *Alice in Wonderland* and *Through the Looking Glass* by Lewis Carroll; **84** Ken Karp; **85** top: Robert Frerck/Odyssey Productions; bottom: Ken Karp; **86** Marvin Collins/Fran Heyl Associates; **88** Rodney Jones/Zenner/Jones; **94** Ken Karp; **95** John Gerlach/Tony Stone Worldwide/Chicago Ltd.; **97** Bob Daemmrich/The Image Works; **98** Ed Robinson/Tom Stack & Associates; **99** Michael Philip Manheim; **100** Hans Reinhard/Bruce Coleman, Inc.; **103** George Dodge/DPI; **105** Oscar Burriel/Latin Stock/Science Photo Library/Photo Researchers, Inc.; **106** Richard Megna/Fundamental Photographs; **107** Dr. E. R. Degginger